The Silken Thread

The Silken Thread

Five Insects and Their Impacts on Human History

Robert N. Wiedenmann
and
J. Ray Fisher

OXFORD
UNIVERSITY PRESS

OXFORD
UNIVERSITY PRESS

Oxford University Press is a department of the University of Oxford. It furthers
the University's objective of excellence in research, scholarship, and education
by publishing worldwide. Oxford is a registered trade mark of Oxford University
Press in the UK and certain other countries.

Published in the United States of America by Oxford University Press
198 Madison Avenue, New York, NY 10016, United States of America.

Library of Congress Cataloging-in-Publication Data
Names: Wiedenmann, R. N., author.
Title: The silken thread : five insects and their impacts on human history /
by Robert N. Wiedenmann and J. Ray Fisher.
Description: New York : Oxford University Press, [2021] |
Includes bibliographical references and index.
Identifiers: LCCN 2021023160 (print) | LCCN 2021023161 (ebook) |
ISBN 9780197555583 (hardback) | ISBN 9780197555606 (epub)
Subjects: LCSH: Insects—History. | Silk Road—History. |
Silkworms—History. | Insect pests—History. | Insects as carriers of
disease—History. | Insects as carriers of plant disease—History.
Classification: LCC QL462.5 .W54 2021 (print) |
LCC QL462.5 (ebook) | DDC 595.709—dc23
LC record available at https://lccn.loc.gov/2021023160
LC ebook record available at https://lccn.loc.gov/2021023161

DOI: 10.1093/oso/9780197555583.001.0001

3 5 7 9 8 6 4 2
Printed by Sheridan Books, Inc., United States of America

We dedicate this book to our children and grandson: Ivy Fisher, Max Fisher, Trevor Simmons, Connor Simmons, Emily Wiedenmann Keane, and Neal Wiedenmann. May you always be an inspiration and never lose your wonder of the world around you.

CONTENTS

List of Figures   xi
List of Table   xv
Preface   xvii
Acknowledgments   xxi

SECTION 1. SILK AND SILKWORM
1. Moth Spit   3
    Silkworms   5
    Breeding for Silk   7
    Agricultural Revolutions   9
    Silk   11
    Sericulture   12
    Silk as Currency   15
    Silk and Other Ancient Fabrics   17

2. The Silk Roads   21
    Nomads, Horses, and Silk   23
    The Royal Roads   25
    The Silk Roads—Hexi Corridor and Tarim Basin   27
    The Silk Roads—Sogdiana   30
    Caravanserais   32
    Samarkand and Bukhara   33
    The Emir, the Ark, and the Pit   35
    Silk, Spices, Religion   37
    The Decline Begins   38
    Mongol Domination   39
    Not the End—Change   41
    Notes   41

3. Silk Goes East and West   43
    Expanding the Roads   45
    Silk Goes East   46
    Silk Goes West   48
    Silk-Making Moves   49
    Silk Intrigue   50

Silk in the Byzantine Empire   52
Silk Goes to Europe   54
The Jacquard Loom and Its Impacts   55
Notes   58

SECTION 2. ORIENTAL RAT FLEA AND THE PLAGUE
4. In Reverse Order—The Third Pandemic First   61
The Beginning   62
Rodents   64
The Third Plague Pandemic   65
Discovery   67
Mechanisms   70
Fleas   72
And More Fleas   73
Plague, Again   75
Explanations, Missed Opportunities   77
Notes   78

5. Not Just the Plague   79
The Dark Ages   80
Justinian and the Byzantine Empire   81
The First Pandemic   83
The Pandemic Continues   84
The Black Death Appears   85
Pathways for the Plague   86
Spread in Europe   88
Two Pandemics   89
Notes   91

6. Sorting Out the Plague   93
Investigating the Plague   94
The Pathogen   96
The Routes   97
Mammal Hosts   98
Was It the Rats?   100
Insect Vectors   101
Other Fleas   102
Was It Fleas?   105
The Story Continues   106
Other Implications   108
Notes   110

7. The Plague, One More Time   112
Attempted Eradication   113
Introduction to Bioweapons   115
Japanese Bioweapons: Diseases and Insects   116

Unit 731   118
Desperate Moves   119
Notes   120

SECTION 3. LICE IN WAR AND PEACE
8. Lice in War and Peace   125
    Introducing the Amazing Louse   125
    An Unexpected Tangent   126
    Body Lice   129
    Jail Fever   130
    Lice and the Great Hunger   131
    Coffin Ships   133
    Lice and Typhus   134
    Lice in War Time   135
    Lessons Learned   137
    Notes   139

SECTION 4. *AEDES AEGYPTI* AND YELLOW FEVER
9. The Bridge Connecting Silkworms to Mosquitos   143
    Sugar   144
    Sugarcane   144
    From Ancient Origins   146
    Portugal, Prince Henry, and Madeira   148
    Infamy   150
    The Slave Trade Begins   151
    Sugar—and Slaves—On the Move   152
    Stealing a People   153
    Voyages, Perils, and History   155
    Notes   158

10. Yellow Fever in the United States   160
    Outbreaks in the Southern United States   162
    "Yellow Jack" Moves to Memphis   164
    Means of Escape   167
    Cause of the Disease   169
    Weaponized Yellow Fever   171
    The Carrier—*Aedes aegypti*   172
    From Africa to the Americas   174
    The State of Knowledge   176
    Understanding   179
    Notes   180

11. The Caribbean, Carlos Finlay, Walter Reed, and Serendipity   182
    Sugar Connects to Slavery   185
    Early Players   186
    A Singular Obsession   188

Unknown, but Connected, Discoveries   189
Short, Deadly War   191
Walter Reed Enters the Fray   192
Back to Finlay   194
"Absolutely Incontrovertible Demonstration"   195
"Silliest Beyond Compare"—But Correct   196

12. William Crawford Gorgas and the Panama Canal   198
Right Person for the Job   199
Eradicate Mosquitos?   201
Circling the Enemy   202
Connecting the Oceans   203
Panama Is Not the Suez   205
Picking Up the Pieces   206
Re-Enter Colonel William Gorgas   207
Mosquitos Meet Their Match   208
The Preposterous and Impossible Plan Worked   210

SECTION 5. WESTERN HONEY BEE

13. Six-Legged Livestock   215
Pollen and Pollination   215
Bees   217
Bee Behavior and Products   220
Origin and Movement   222
Beekeeping Begins   223
Ancient Beekeeping   224
Float Like a Butterfly, Sting Like a Bee   227
Expansion to the West   228
Importance of Bees to Almonds   229
Bees in Trouble   231
Notes   232

SECTION 6. TYING THE SILKEN THREADS

14. Tying the Silken Threads   237
Impacts—Measurable and Immeasurable   237
History That Is Not Found in History Books   238
Egypt, Syria, and Fleas   240
Haiti, Slave Revolt, and Yellow Fever   242
The Cold Winter and Lice   243
Never Doubt for a Moment   245

Bibliography   247
Index   259

FIGURES

1.1. Adult silk moth (*Bombyx mori*) on a curved branch.   4

1.2. Adult wild silk moth (*Bombyx mandarina*).   5

1.3. Adult female silk moth (*Bombyx mori*) and wild silk moth (*Bombyx mandarina*).   7

1.4. Silkworm (*Bombyx mori*) caterpillars on tray of white mulberry (*Morus alba*) leaves.   13

1.5. Silkworm cocoons on trays.   15

1.6. Collecting silken threads from silkworm cocoons.   16

2.1. Satellite image of the Tarim Basin, Xinjiang, China; with the Tien Shan Range to the north (top), the Kunlun Shan Mountains to the south, and the Pamir Mountains to the west. Note the enormous dust storm in the eastern Taklamakan Desert, which occupies most of the basin.   22

2.2. Major routes that comprised the Silk Roads spanning Asia.   24

2.3. Routes that comprised the central core of the Silk Roads in western China and Central Asia.   29

2.4. Flaming Mountains in the Tarim Basin and the edge of the Taklamakan Desert, near Turpan, Xinjiang, China.   30

2.5. Snow-capped peaks of the Hindu Kush Mountains. Some routes of the Silk Roads traversed passes through the Hindu Kush.   31

2.6. Caravanserai in the desert, along a Silk Roads' route.   33

2.7. Geometrical architecture and mosaic art, Samarkand, Uzbekistan.   34

2.8. Ulugh Beg Observatory, Samarkand, Uzbekistan.   34

2.9. Ancient Muslim architectural complex, Ark fortress, Bukhara, Uzbekistan; a UNESCO World Heritage Site.   36

2.10. The tomb of the Asian conqueror Timur, in Samarkand, Uzbekistan.   40

3.1 Maritime Routes that complemented the land-based Silk Roads.   44

3.2. Persian silks displayed in a market along the Silk Road in Uzbekistan, Central Asia.   49

3.3. Jacquard loom showing chain of punch cards.   56

3.4. Herman Hollerith (1888), inventor of punch card and reader system. *57*

4.1. South Asia and Southwest China and the Tibetan Plateau. *63*

4.2. Great gerbil (*Rhombomys opimus*); Muyunkum desert of South Kazakhstan. *65*

4.3. Dr. Alexandre Yersin, co-discoverer of the plague bacterium. *68*

4.4. Dr. Shibasaburo Kitasato, co-discoverer of the plague bacterium, *Yersinia pestis*. *69*

4.5. Dr. Alexandre Yersin in 1894, in front of his straw hut laboratory in Hong Kong, where he first isolated and described *Pasturella pestis*, now known as *Yersinia pestis*. *70*

4.6. Left lateral view of a female Oriental rat flea, *Xenopsylla cheopis*. *73*

4.7. Interior of a temporary hospital for plague victims, Bombay epidemic, 1896–1897. *76*

5.1. *Plague at Elliant* by John Everett Millais, engraved by Joseph Swain. *90*

7.1. Workers in the southern Soviet Union putting poison in rodent burrows. This photograph was taken sometime between 1930–1940. *114*

7.2. Building on the site of the bioweapon facility of Unit 731, near Harbin, China. *117*

8.1. Female body louse, *Pediculus humanus*. *127*

8.2. US soldier demonstrating hand-spraying equipment used to apply DDT. *138*

9.1. Cut sugarcane stalks. *147*

9.2. Madeira Island terrain. Terraces used to level ground for cultivation are visible. *149*

9.3. Slavery in Africa. The Treaty, vintage engraved illustration. Journal des Voyage, Travel Journal, (1880–81). *155*

9.4. Major trans-Atlantic slave routes; from Angola and the Slave Coast to western terminuses in Brazil, the West Indies, and the United States. *157*

10.1. Major trans-Atlantic trade routes in the mid-1700s, forming the three legs of the First, Middle, and Final Passages. *161*

10.2. Southern United States and the Mississippi Valley, showing proximity to Havana, Cuba. The bold line represents the short route of the Memphis & Little Rock Railroad. *163*

10.3. Bird's eye view of the city of Memphis, Tennessee, 1870. *165*

10.4. Burying yellow fever victims during the 1878 Memphis yellow fever epidemic; Elmwood Cemetery. *167*

10.5. Yellow fever mosquito, *Aedes aegypti*; ovipositing eggs, which are laid individually on a vertical surface, not on water. *173*

10.6. Dr. Carlos Finlay, Cuban physician; his experiments were instrumental in understanding the transmission of yellow fever. *177*

10.7. Major Walter Reed. *178*

11.1. The Caribbean Sea, Caribbean Islands, and Central America. *184*

11.2. Brigadier General George M. Sternberg, US Army Surgeon General. *188*

11.3. Theodore Roosevelt and Rough Rider officers during the Spanish-American War. Colonel Roosevelt is in the front row, fifth from left. *192*

12.1. US Army Surgeon General William Crawford Gorgas in 1914. Gorgas was responsible for eliminating yellow fever in the Panama Canal Zone. *200*

12.2. Ancon Hospital grounds, Panama Canal Zone, ca. 1910. *209*

12.3. President Theodore Roosevelt running an American steam-shovel, Panama Canal, 1906. *210*

12.4. Construction on the Pedro Miguel Locks, part of the Panama Canal. *211*

13.1. European (western) honey bee, *Apis mellifera*; note pollen basket on hind leg. *218*

13.2. Black-eyed Susan in visible (to humans) light (left) and in ultraviolet light (right); note the "nectar guides" that may lead honey bees to the flower's pollen source. *221*

13.3. Bee hives for pollination in an almond orchard, Central Valley, California. *230*

14.1. *Retreat from Russia*, by Pryanishnikov. *245*

TABLE

2.1. Dynasties and empires, their primary periods of activity, and the regions (modern-day) they occupied during the time frame of the Silk Roads.   *26*

PREFACE

ZILLIONS

"Zillions" is the technical term of choice that specialists, like us, use when referring to the awe-inspiring diversity of insects. And yet, if there are zillions, why just five? Why not six or fifty? Or more?

FIVE MILLION

With about a million species already known, insects account for more than half of all *known* life. But even conservative estimates suggest there are more than five million insects, including those yet to be discovered (compared with an estimated 500 more mammals yet to be discovered). Five million insects, five million stories. Collectively, those five million stories of success illustrate a different world than you may normally observe. Insects don't just succeed on this planet—they thrive and dominate. Insects do not inhabit our world—we inhabit theirs. Let us convince you of this, not with the collective weight of five million stories, but with the stories of five. Nay, not even the complete stories of five insects, but only the parts that have dramatically altered aspects of our own stories.

FIVE INSECTS

Why five and why these five? We chose these five because of our perception of their impacts; others may disagree with the five we include, but we stand by our choices. As to why five and not eight or ten or thirty? Where do we start and where do we stop? Five insects—these five— allowed us to tell some history and slip in a little entomology without turning you into a bug-nerd.

The task of narrowing the list to five insects that have impacted human history may seem simple. For example, which insect-borne diseases result in the greatest death toll? Easy. Malaria has caused—and still causes—millions of infections and hundreds of thousands of deaths each year, so the mosquitos that transmit malaria are an obvious choice, right? Actually, no, for several reasons. For one thing, "malaria" is not a single entity. Multiple species of the pathogen (*Plasmodium*) cause malaria (five species infect humans, but over 200 species of *Plasmodium* are known). Also, it isn't about one mosquito either. Malaria is vectored by mosquitos of a single genus, *Anopheles*. But there are 460 known species of this genus, with 30–40 species that commonly transmit the pathogen. Each of them has its own story. Make no mistake—we recognize malaria as one of the worst scourges affecting humans, but the insect part of the story is not simply told. It would take an entire book to explore all the mosquitos that transmit diseases, and this is not that book.

We have tried to peer beyond death tolls and select five insects at key junctures in history, the convergences of place and time, and show how they shaped the lives of historical figures or how they caused or were linked with important events in human history. Our history. The history that we thought we knew well. It turns out we didn't. Actually, we never learned about insects' roles in history to begin with—maybe it's the same for you. As we dug a little deeper, it turns out all five insects can be linked together by the Silk Roads, either directly, by trade routes or diseases, or indirectly, through paths that at first seem arcane but make sense when brought to light. Hence, the title. We try to keep that silken thread prominent throughout the book.

THE LIST

In our list of five insects, three have been responsible for transmitting horrible diseases that have sharply altered history. These include yellow fever mosquitos (*Aedes aegypti*), Oriental rat fleas (*Xenopsylla cheopis*), and body lice (*Pediculus humanus*). All three are vectors of diseases that have killed more humans than all the wars in history. That is a staggering statistic, to put it mildly. Each of them still, today, spreads the same diseases that have caused unimaginable suffering. However, realize that those three insects have not killed anyone: none, nada, zip, zero. The pathogens they vector are the killers.

For our fourth species, we chose an insect that is revered rather than reviled. Silkworms (*Bombyx mori*) are grown and harvested for a product that we cherish. The story of silk is about more than silkworms—it is also the story of humans. Silkworms played a major role in ancient history. Domestication of plants and animals led to local surpluses that were traded or sold in trade centers, connected in a network spread across Asia, a network of roads called the Silk Roads. Trading along those roads involved more than silk. The exchange of goods, ideas, and religions along the Silk Roads created a history that continues today, still vibrant.

The last of our five insects, the Western honey bee (*Apis mellifera*), is both revered *and* reviled. This species is well known. Perhaps you have kept honey bees yourself or have seen bee yards with their hives, or at least you noticed worker bees foraging at flowers or picnic areas. This familiarity belies the importance this species has played in our history. Honey bees have altered agriculture and enabled the diversity of foods that we eat. On the other hand, they are farm animals, not wildlife. As such, they can harm native environments and species. In the case of honey bees, the primary concern is harm to native bees. Add to that the concern that a bee's sting that can cause life-ending anaphylactic reactions. Indeed, honey bee stings kill more people in the United States than the bites of snakes and spiders, which many people fear much more than bees. Honey bees—revered *and* reviled.

THE BOOK

This book began as periodic lunch conversations between the two of us. In fact, it started with a story about the Memphis & Little Rock Railroad and the idea of quarantining passengers for 21 days (Chapter 10) taught by RNW in a course ("Insects, Science, and Society") at the University of Arkansas. A 21-day quarantine seemed foreign to us when we began writing this book. Since then, the COVID-19 pandemic hit and terms like "quarantine" and "self-quarantine" have become part of everyday language.

We knew we could not write a book telling the stories of the zillions or, worse, had we tried, we would have produced something of interest only to other entomologists. That was not our intent. Instead, as we discussed certain insects or historical events that had been featured in class lectures, we realized in amazement how even just a few insects had affected history. JRF came up with the idea of narrowing the zillions to just five, but to tell the stories and history behind those five. As we explored a little further, we realized that much of that history was unknown to us, too. The more we explored, the more vibrant the stories became. We tried to put ourselves into the reader's shoes (or armchair), and we tried to assume the perspective of a reader, not of an entomologist. We have tried to keep jargon to a minimum and to explain some entomological processes, anatomy, and interactions in language understandable to all. We did not "dumb down" the information, we just tried to make it accessible.

Undoubtedly, the chapters contain more history than entomology. Some of that history-intensive writing is to give the historical "back story," some is to illustrate the web of impacts insects have had on us—in ways that will likely be surprising to you.

History is written from the perspective of winners, but maybe we don't realize that we were not the winners. Humans bask in their success, unaware that it is due largely to organisms less than two centimeters long. Few chronicles of the history of *Homo sapiens* even acknowledge insects' influence. But, insects have not just influenced human history, they have driven it. They still do. We maintain that the jury is out on who the winners in history will be, but our money is on the insects. We try to tell a little of that history, the history that failed to make it into our textbooks.

There is an adage, "When the only tool you have is a hammer, everything looks like a nail." We tried to keep that in mind, recognizing that not every event in history was affected by insects. We are sure there were a few that were not. We can't name them now, but we are sure it is true. Well, pretty sure. OK, OK, hammers and nails . . . .

DISCLAIMERS

First and foremost, we are not historians; we are entomologists with a love for history. We did not attempt to tell the complete history of Eurasia, Chinese Dynasties, the Byzantine Empires, or Napoleonic Wars. Plenty of history books do that more completely than we could. Because of our backgrounds and interests in the things that count, our view of history will differ from the perspective of bona fide historians. They no doubt know their stuff. But they also usually overlook what renowned entomologist E. O. Wilson called "the little things that run the world."

Second, some of the stories are still unfolding, being rewritten as we were writing. No doubt the complete stories will be different a decade from now. We tried to represent the best of current knowledge, or at least our interpretation of that knowledge.

And third, we have tried to avoid being US-centric. We are both from the United States, and that forms most of our experience base and perspective, even though both of us are well-traveled and attuned to the rest of the world. Despite our efforts and intent, we likely failed to make the perspective broad enough. Still, we tried.

CAVEAT

We need a quick caveat here. One of the most fascinating aspects of biology, that sets the field apart from some other academic fields, such as physics and mathematics, is that there are numerous exceptions to nearly everything biological. Be wary when any authors use words like *all*, *every*, or the worst one, *never*. Alarm bells should be going off whenever you see those words, and we tried to avoid them and, instead, offered wiggle words, such as *except for*, *in most cases*, or *nearly always*. Alas, in biology, rules were made to be broken. Can you imagine if mathematics and physics operated this way? Something like: the Pythagorean theorem works in all cases, except for a few special triangles that occur on small islands in the Coral Sea; or Newton's law of gravitation sometimes gets suspended when there are five Thursdays in a month. Well, exceptions in biology are not always exceptions, especially with insects. So, keep the caveat in mind as we offer the descriptions that follow. Except when . . . .

We hope that these stories fill out the history you know, or that our telling of these stories may cause you to re-think published history and, instead, include in your perspective the ones that get our votes for the real winners, those zillions of six-legged hammers looking for nails.

ACKNOWLEDGMENTS

This book began as a series of lunchtime conversations between the two of us, gleaning stories from lectures given in classes we both taught at the University of Arkansas. The project was aided by the generous help of key people along the way. Jeff Lockwood shared helpful information on book proposals to get us started. Gene Kritsky offered early support and introduced us to our editor at Oxford University Press, Jeremy Lewis. Jeremy led us along the publishing path, giving us free rein when possible but handholding when necessary. His editorial assistant Bronwyn Geyer helped with tracking down photos, permissions, seeing to the necessary forms and helping guide us in the later stages of production. We benefited greatly by the skills and insight of copy editor, Bríd Nowlan, and Production Editor, M. Ponneelan. We were encouraged by Luann Wiedenmann, S. Raghu, Tim Kring, Jim Maloch, Fred Stephen, and Max Meisch; Jim, Max, and Fred read early drafts of the chapter on yellow fever in the United States. We extend our thanks to Hammontree's. Matt Bertone and Andrew Davidhazy kindly provided photographs. Sarah Silva of MapShop.com created and revised (again and again) the maps, which added greatly to the final project. Students at the University of Arkansas were a joy to teach and to share the subject matter with.

We owe an enormous debt that can never be repaid to our dear friend, Professor Gimme Walter, who not only offered encouragement but read every word of the chapter drafts, found the flaws in logic and missing connections, offered alternative wording that we accepted, and added some wonderfully dry humor to the text. Gimme—Good on 'ya, mate! Any errors of omission or commission are ours, not Gimme's.

SECTION 1

Silk and Silkworm

—6000 BCE Chinese Agricultural Revolution
—5000 BCE Domestication of silkworm begins
—206 BCE Han Dynasty begins
—119 BCE Silk Roads begins (informally)
—1 CE Common Era begins
—476 CE Western Roman Empire falls, Eastern (Byzantine) Roman Empire persists
—560 CE Constantinople and Byzantine Emperor Justinian obtain silkmaking skills
—980 CE Ibn Sina (Avicenna) is born
—1220 CE Genghis Khan founds Mongol Empire
—1275 CE Marco Polo arrives at the court of Kublai Khan
—1370 CE Reign of Timur and Timurid Empire begin
—1405 CE Silk Roads begin to decline with the death of Timur
—1453 CE Byzantine Empire falls
—1775 CE Jacquard loom, using punch cards, is invented in France
—1807 CE Bassi begins study of silkworm disease
—1831 CE Silk worker uprising in Lyon
—1838 CE Charles Stoddart is imprisoned in the Ark fortress by the Emir of Bukhara
—1896 CE Herman Hollerith begins the Tabulating Machine Company, using punch cards

CHAPTER 1

cᐯɔ

Moth Spit

Moth spit. That is hardly a glamorous descriptor for the robes and gowns of emperors and empresses. Nonetheless, spit from the young of certain moths has captivated humans for centuries. Although the moths are just doing what they have done for eons—secreting proteins from their mouths to spin into cocoons as part of their life cycle—colloquially, we call that moth spit by a more refined name: silk.

Human production of silk is so ancient it is difficult to state with any certainty just how old it is, with competing legends, tales, and other evidence that nevertheless offer some insight as to its age. As with most apocryphal tales, the story told most often of silk's discovery may have a grain of truth or, in this case, a thread of truth. According to this predominant legend, nearly 4,700 years ago the Empress Lady Hsi-Ling-Shih, wife of Huangdi, the mythical Yellow Ruler, was sitting under a mulberry tree having tea, when an object fell into her teacup. When she pulled the white oval object from her cup, she found that it was made up of a sticky filament, and as she pulled on that thread, its lustrous sheen captured her attention.

A similar legend tells of when Huangdi, leader of China's northern people defeated the leader of the southern tribe. Having seen the power of Huangdi, the "Goddess of Silk" presented him with a ball of silk thread spun from "mulberry cocoons." Huangdi ordered the ball of silk to be woven into cloth, which was then presented as a gift to his wife, the Empress Lady Hsi-Ling. She found the garment so pleasing that she collected the mulberry cocoons herself, and the weaving of silk ensued.

In both myths, the empress discovered that unraveled thread from the white, lightweight object could be woven and made into fabric, ultimately leading to the practice of sericulture, the culture of silkworms, and harvesting of silk. The legend also claims that she invented the loom to allow the use the newfound silk thread. Her husband, Huangdi, was credited with introducing a form of government and the use of coins as money for trade during his reign and for spreading the virtues of Taoism throughout his empire. Wait a minute—they're both mythical, right? Attributing so

The Silken Thread. Robert N. Wiedenmann and J. Ray Fisher, Oxford University Press. © Oxford University Press 2021.
DOI: 10.1093/oso/9780197555583.003.0001

many concrete examples of inventions or introductions to two mythical figures is hard to understand. Even if the tales are apocryphal, what a power couple!

Regardless how the discovery was made, or exactly when, silk in antiquity was the fabric of the emperor, made for and worn exclusively by the imperial families in China. The emperor supposedly wore white silk robes when he was inside the palace but outdoors, or away from the palace, he wore yellow robes—yellow symbolized the color of the earth.

The stories of Empress Hsi-Ling-Shih are just a few of the tales told to explain the origin and use of silk. Like many tales, they are not true in their current form. Both the Yellow Emperor and his empress were mythical figures. The legends at least point toward an ancient recognition of silk and its possible uses, but even the mythical figures do not go far enough back to have been silk's originators. Some archaeological records date the origin of silk production nearly 2,000 years earlier than the legends of the Empress Lady Hsi-Ling. Pottery depicting silkworms and spinning, spinning tools and looms, and even silk fabric itself, has been dated to at least 5,000 to 7,000 years ago.

Part of the difficulty in nailing down a date is that domestic silk moths—*Bombyx mori* (Figure 1.1)—are not the first or only species to yield silk. Wild silk moths—*Bombyx mandarina* (Figure 1.2)—were domesticated through years of selective breeding to create domestic silk moths. The process of domestication began more than 7,000 years ago—within the time frame of the archaeological relics, but well before the empress' tea cup. The selective breeding that produced domestic silk moths lasted

Figure 1.1 Adult silk moth (*Bombyx mori*) on a curved branch.
Source: Ozgur Kerem Bulur/Shutterstock.com

Figure 1.2 Adult wild silk moth (*Bombyx mandarina*).
Source: DeRebus/Shutterstock.com

for two millennia, which would put the earliest date for complete domestication in the neighborhood of 5,000 years ago.

Historical records do not reveal just when the harvest of silk from *Bombyx mandarina* became the harvest of silk from *Bombyx mori*. Robust genetic evidence and archaeological finds may push back the origin of silk, but those records lack the glamor of a legendary empress or a close association with prominent mythical royalty. As a result, the legend persists and is even manifest in structures, not just tales. A building in Beijing's Beihai Park, near the imperial compound now called the Forbidden City, is home to the Hall of the Imperial Silkworm and the Altar of Silkworms, where mulberry leaves were inspected before being fed to the imperial silkworms, which were grown in small structures along walls of the hall. So, tale or no tale, silk was taken seriously enough to warrant buildings named in the moths' honor.

SILKWORMS

What is this thing called silk? To answer that, we need to consider the producer, silkworm caterpillars. Domestic silk moths, *Bombyx mori*, are classified as Lepidoptera, the insect group comprising moths (butterflies are classified as a grouping within moths). The larval stages of Lepidoptera are typically given a special term—caterpillar—but the caterpillars of many, especially pest species, are often referred to as "worms." Case in point, the caterpillars of domestic silk moths are called silkworms; in fact, the species itself is often referred to as silkworms rather than "domestic silk moths." We use these interchangeably. During development, silkworm moths undergo complete metamorphosis, meaning they have a life cycle consisting of an egg stage, several larval stages (five, in the case of silkworms), a pupal stage, and then they emerge from pupae as adult moths. Silkworm moths have been selectively bred for thousands of years to yield a species that produces abundant and high-quality silk. However, silkworm moths were not always like that.

Wild silk moths, *Bombyx mandarina*, are the ancestors of contemporary silk moths and are one of the eight species of *Bombyx*, a genus that is collectively referred to as "true silk moths." Wild silk moths are found in eastern Asia, including China, eastern Russia, Korea, and Japan, where this species overlaps in distribution with other true silk moths. *Bombyx mandarina* also produces fine-quality silk, which undoubtedly was noticed thousands of years ago by locals in China, thus, the cocoon in the empress' tea cup.

At some point, thousands of years ago in central China east of the Tibetan Plateau, someone recognized that the silk produced by northern populations of *Bombyx mandarina*, being finer than silk produced by other local silk moths, had potential as a product. Even more important was the recognition that selecting the best individuals for breeding could improve both the amount and quality of the silk produced. We humans are masters of this process, called domestication, and have engaged in partnership with many other species, but most of them—such as pigs, chickens, and rice—were grown as food. Silkworms, in contrast, were grown for a luxury product. By choosing, for breeding, only those individuals with certain desired qualities, silkworm farmers improved the quality of the silk they produced. So began another of humankind's successful partnerships with other species, and the results are remarkable (Figure 1.3).

We need to pause for a moment here. We are suggesting this was one of the most successful breeding programs ever? Yes. And it began 7,000 years ago? Again, yes. And we are further suggesting that this tremendous effort required thousands of years? For a moth? Surprising, but yes. What seems most remarkable is that it is not clear that the selective breeding of silk moths was under the direction of any one person. It was not that someone had a vision of just how valuable silk could be and then set about to undertake the effort that ultimately took thousands of years. This commitment seems like something an imperial ruler could order, but that is not clear from historical accounts. Although silk production was widespread through a number of Chinese provinces, breeding better moths does not appear to have been a monolithic, coordinated effort. Countless individual silk farmers grew silk moths, cultivating them to produce fine-quality silk, which could be sold—a proximal outcome. Presumably better silk fetched a better price, so there was an incentive for each silk moth farmer to produce the best silk possible. There was no way any of the silkworm farmers knew in advance that breeding would result in a domesticated moth population producing highly prized silk far into the future. Not only did they not know this, they likely did not even consider such an ultimate outcome.

Figure 1.3 Adult female silk moth (*Bombyx mori*) and wild silk moth (*Bombyx mandarina*).
Source: Courtesy Markus Knaden, Max Planck Instutute for Chemical Ecology

BREEDING FOR SILK

Pause again for a moment and put these breeding experiments into context. Nothing was known about genetics—not genes, chromosomes, or modes of inheritance. It would be more than 6,000 years before any of that would be known. What tools for improving moths by selective breeding did the silkworm farmers have at their disposal 7,000 or more years ago? One tool they had was the most important one of all, as any insightful scientist will attest—skill at careful observation. Breeding for selected characteristics required observing the outcome and keeping records, whether written or oral. Despite not understanding the process, observant silkworm farmers had in their favor the very same two factors that support evolutionary changes: numbers and time.

Variability occurs in all organisms. Look at people around you: tall, short; different colors of eyes and skin. These characteristics are controlled by our genetic makeup, by genes that are found on chromosomes. Some variability is due to new combinations of genes. Still other variability is due to mutations during the reproductive process. For sexually reproducing organisms, the union of gametes and recombination of chromosomes sets the stage for mutations to occur. Mutations happen naturally; sexual reproduction isn't always neat and predictable. Any one mutation may be rare, but mutations occur all the time.

Mutations can be deleterious, neutral, or advantageous. Deleterious mutations often result in death or seriously impact life—think about some of the diseases affecting humans that occur as a result of a genetic mutation. In contrast, advantageous mutations confer some advantage—bigger, stronger, more colorful, or a host of other traits. And some mutations are neither advantageous or deleterious; we term them "neutral." They may not benefit the organism, but at least they do no harm—sort of a Hippocratic Oath among genetic mutations.

The process of natural selection puts any mutation to the ultimate test: reproductive success. Does the organism with the new trait produce more offspring, or at least not fewer offspring? Long stretches of time provide opportunities for those individuals with a beneficial mutation to produce more offspring with that new trait in each generation. The outcome, greatly simplified, is evolution by natural selection.

The processes of natural selection also operate when humans selectively manipulate the reproduction of animals and plants to produce an outcome desired for one reason or another. The same variability that occurs from naturally occurring mutations is at the heart of human breeding of organisms. If there is no variability, there can be no selective breeding. In the case of silkworms, large numbers of silk moths were grown by many individual silkworm farmers. The large numbers provided ample opportunity for variability to occur and be recognized. With silkworms, the variability did not yield greater reproductive success, as that aspect was influenced by the growers' breeding regimens. Instead, the variability that affected silk itself was what breeders recognized. Selective breeding allowed the production of "strains" of silk moths with different qualities, and some may have produced more silk or silk of a better quality. Large numbers of silkworms grown by numerous farmers provided part of the fodder for moth breeding.

The other factor that benefited the selective breeding of silkworms was time. It takes many generations for accruing mutations in an organism's genes to cause the physical changes often desired by selective breeding. Two seemingly opposite aspects of time played important roles in silkworm domestication: long time and short time. Long time represented thousands of years, a small amount of time in evolutionary terms, but a nonetheless a timespan needed for evolutionary change. Short time was the duration of the insect's life cycle, in the case of silkworms, less than a month. A short life cycle acted as a generational multiplier to each of the thousands of years, which meant even more opportunities for mutations to occur.

There was also a human element to silkworm domestication. Beneficial changes in moths were not universally found by all the silk moth breeders. The large numbers of breeders and larger number of moths would have yielded, at some point, offsprings that were clearly better than their predecessors. But for that "improved" line of moths to become more numerous, there needed to be lateral spread among farmers—the human aspect—involving the sharing or trading moths with desired traits. Gradually, the "improved" moths would have become widespread as the newer moths were more desirable to farmers than their ancestors.

Wild silk moths are distributed across a broad geographic range, across China and Mongolia into Korea and Japan. Traditionally, it was thought that wild silk moths were divided into two groups, one on mainland Asia and the other in Japan. We now know the important divider may actually be the Qinling-Huaihe Line (named for the Qinling Mountains and the Huai River). This geographical feature divides the region into a tropical southern region and a cold, arid northern region. It also divides

Chinese peoples into northern and southern in terms of culture, cuisine, and life-style. So too does it divide northern and southern wild silk moths. Of moths and men. Southern populations of wild silk moths occur in both the southern half of China and Japan. Those moths that partnered with humans more than 5,000 years ago to become domestic silkworms came from populations in Northern China. Some researchers suspect the Qinba Mountains could be their birthplace. For now, we must be satisfied that in the drier, colder north of China, east of the Tibetan Plateau, Stone Age farmers looked to local wild silk moths and thought, maybe, just maybe, they could wear that moth's spit.

Domestic silk moths bear little resemblance to their ancestor, in part because of something called pleiotropic effects, meaning multiple effects from one change. All genes have multiple effects, so changing one thing can also change another. Many genes are linked to other genes, so choosing one trait can lead to another accompanying genetic change. As silk moth farmers chose moths that produced one desired outcome, such as more silk, other characteristics came along for the ride, such as the color of the hair (called setae). Pleiotropy and linked genes explain why selection on silk quality or quantity eventually produced moths that are big, bulky and clumsy, blind, covered in white hairs, and with wings that are tiny vestiges of their former selves, all of which contribute to the moths being unable to fly. Add to those traits the inability to reproduce effectively without human help. Domestic silk moths are almost totally reliant on continual human care. With all the traits that were lost as the moths were bred for better silk, adults can no longer even do stereotypical "moth things"—they cannot even fly into a candle.

Some of the traits selected in silkworm breeding were beneficial to the farmers. Flightless moths could not escape, and the moth's need for assistance with mating meant farmers completely controlled which individuals mated, thus hastening the effects of breeding for desired traits. Someone at the time also understood the need to periodically interbreed the new moth line with its ancestor, *Bombyx mandarina*, to produce healthy offspring that were not affected by inbreeding. Again, recognize that all this occurred thousands of years before there was any understanding of genetics. And none of this massive breeding program is mythical—it is the real deal.

AGRICULTURAL REVOLUTIONS

Selective breeding of silkworm began toward the latter years of the Chinese Agricultural Revolution. Where does that stand in the history of agriculture? Somewhat earlier in history, the Neolithic Agricultural Revolution represented the transition from a nomadic lifestyle to a more sedentary way of life. Estimates of the dates for the Neolithic Revolution are about 12,000 years ago, beginning in south-west Asia and lasting for nearly 6,000 years. As populations of humans settled down during the Neolithic Revolution, they began to domesticate plants, such as barley, wheat, and lentils, and animals, such as sheep, goats, and cattle, which provided food.

The Chinese Agricultural Revolution began slightly later and independent of the agrarian changes happening in Western Asia. The plants and animals domesticated in China differed from those species in Western Asia, too: millet, rice, pigs, chickens, and geese, for example. At that point, all the domesticated plants and animals, whether in Western or Eastern Asia, were food species, and their domestication led to more reliable food sources, which enabled, supported, and *demanded* the transition of people to a sedentary lifestyle.

Another agrarian revolution occurred with the recognition that domesticated animals could provide other products: eggs, milk, manure, and power as draft animals. These products did not require "harvesting" the animal but recognized that the animals were useful on an ongoing basis. This second agrarian revolution with domesticated animals still yielded products of proximal utility: for the here and now.

Still one more agricultural revolution occurred, termed the "Secondary Products Revolution." This last revolution was much more complex than we have time and space to go into here, but it figures prominently in our discussion about silk and silkworms. The Secondary Products agricultural revolution came about from the knowledge that animal or plant products had value combined with the recognition that the agricultural products grown as food also provided the raw material for producing a refined product, or a craft good.

Refining a product to a craft good increases its value, in part by increasing its longevity. Food has immediate value to the farmer but, in the longer term, food can spoil, sometimes very quickly, thus limiting its value. Again, its value is here and now. However, a surplus agricultural product that is converted to a crafted product lasts longer, can be moved around readily, and maintains its value for trade or barter.

Domestication of silkworms fits the secondary products movement, but it is also unique in several ways. First, the product of silkworm domestication definitely was the raw material for producing a craft good, such as a silk robe. However, unlike other animal products in the Secondary Products Revolution, silk production did not really have an intermediate step, because the animal that produced silk did not have other uses or value. Second, the domesticated silk producer was a small insect, not a large mammal or bird, which required constant attention and had specialized food requirements. And third, there was no immediate benefit to silkworm breeders right there and at that time—silk could not be consumed, nor did it protect the user from the elements. Silk was a luxury product. This had far-reaching consequences, as you will see.

The raw material that silkworms yielded, once converted to a refined product, could be exchanged and transferred to others over great distances and could last a long time—a craft good whose value was no longer limited to here and now. Products from other domesticated species could be traded, and they certainly were, but silk had different kinds of utility. It didn't spoil, as food products did. Further, although silk's production required very demanding husbandry, silk itself didn't require care and husbandry en route to or from trading events, as cattle or sheep did.

Silk domestication occurred over thousands of years. Because its utility spanned space and time, silkworm domestication might be considered to have accelerated the emergence of modern humans. That change in utility of a domesticated species reflected more than a transition from pastoralism to agrarianism. The domestication of silkworms provided craft goods that could expand interactions among humans.

SILK

Thousands of years were spent in selective breeding for better moth spit? Yes. What exactly is in moth spit that makes it silk? Proteins. Silk from *Bombyx mori* is made of two kinds of proteins, fibroin and sericin. The silky spit extruded from the salivary glands consists of two filamentous strands of fibroin, covered with a layer of sericin that acts as an adhesive. Sericin, the sticky protein that holds the fibroin strands together, is stripped away when the cocoon is immersed in boiling water or treated with steam (or, perhaps, dropped in a tea cup). The remaining protein, fibroin, is made of a sequence of just three different amino acids—glycine, serine, and alanine, arrayed end to end like railroad freight cars—and the sequence of those amino acids is what gives it strength.

Silk has high tensile strength, with the often-cited statistic that silk is stronger for its weight than steel. Tensile strength is a measure of how much stress can be applied to a material before that material breaks. Engineers say that silk is equal to or stronger than steel. True that, but how many suspension bridges are made with silk? Maybe not all that many. In fact, none. However, that same tensile strength also makes silk a valuable material for clothing manufacture, because silk can tolerate a lot of stretching without breaking. Silk's tensile strength is also ideal for protective gear, such as military combat helmets or bullet-proof vests: it is strong but lightweight and not uncomfortably hot when worn. Silk is hypoallergenic and, as well as being strong, it is soft enough to alleviate skin-irritating conditions so is used as a wound dressing; its moisture-facilitating ability also enhances healing. Strong, yet soft. It seems that the simple chemical composition of the fibroin protein produced by our clumsy, flightless, blind silk moths is very special indeed.

Silk is prized for its luster, sheen, and beauty, which accounts for its use in fine garments, such as robes and gowns. The structure of the fibers gives silk its sheen, as it reflects and refracts light from different angles, so that the woven fabric changes color and glows vividly. Because of silk's beauty, its use for clothing was restricted to the emperor and his family for centuries after the production of silk began.

At some point, permission to wear silk was extended to high-ranking officials and wealthy merchants. The wealthiest established their own silk production "factories," vertically integrated from the harvesting of mulberry leaves and feeding of silkworm larvae, to the weaving of the harvested and spun silk into fabric. These factories employed hundreds of women working in dedicated workshops, where they produced enough silk to meet the needs of the estate, as well as excess silk to sell.

Wearing silk was still limited to those of high social rank, whether by imperial decree or by the exorbitant cost of silk clothing. The finest silk, dyed and embroidered, was a symbol of social status. Only in later years, and after production expanded into Korea at about 200 BCE, with a large emigration from China, did the wearing of silk extend to the lower classes, including the farmers who produced it. Even then, commoners were restricted to wearing plain silk clothing.

Silk use went through several transformations in Early China: beginning as a highly exclusive fabric reserved for the imperial family and becoming a surface to be used for decorating, such as on hand fans or wall hangings. It also became the surface used for painting portraits, landscapes, and scenes from well-known poems and tales. Series of paintings about tilling and weaving were reminders of the contribution of China's agriculture to prosperity. The theme of the paintings aligned with the traditional division of labor at the time: "men plow, women weave."

Silk was amenable to dyes and inks made from natural materials, and Chinese artists of the first millennium BCE were highly skilled in producing works of art. At that time, their silk art involved the use of dyes made from minerals, such as cinnabar and red ochre, or inks like indigo, derived from plants. As production became more sophisticated, the finest silks reserved for the emperor were augmented with embroidery and patterns. High-quality silks were also used to pay tribute to an emperor, or as gifts from the emperors that symbolized his benevolence. In just one year of his reign, the Han emperor Wu gave gifts amounting to 20,000 rolls of silk.

The transformation of silk, from clothing to art surface to utilitarian good, provided a significant economic boost to the empire. Products such as fishing lines and nets, bowls to transport water, musical instruments, string for archery bows, and high-quality writing paper provided new and expanded uses for silk.

Silk's beauty was and is still the basis for its use as fabric, but its qualities went far beyond beauty and its uses went beyond fabric, to include art and industrial products. One reason for silk's timeless value is that it is impervious to the growth of damaging fungi and attack by most species of bacteria. Silk is also not fed upon by that bane of clothes hung in closets everywhere—clothes moths. There is a good reason why silk is safe from the feeding of clothes moths, or so renowned entomologist May Berenbaum claims, with her inimitable humor: "I believe it is professional courtesy." It is a "courtesy" accorded by clothes moths to the silk products made by a kindred spirit—another moth.

SERICULTURE

For millennia, silk was made solely in China. The ancient practice of producing silk, called sericulture, began thousands of years ago and, in many regards, is still the same today. Two factors were most important: the diet fed to the larvae and the prevention of complete development by interrupting the pupal stage of the moth. The early Chinese perfected those, and the entire process has remained mostly unchanged.

Sericulture begins with growing and harvesting leaves from the white mulberry tree (*Morus alba*), as apparently the best silk is derived from silkworms that feed on white mulberry. Due to the need for frequent harvesting of leaves over half of the year, mulberry trees are located near homes where silkworms are grown, in a practice both ancient and contemporary.

Silk's production was distributed among households as a sort of cottage industry, creating a unique production economy. Because the silk industry was so important for family income, the idea was advocated that all land holdings, even the very smallest, should have a small area on which mulberries were planted. Further, laws were enacted to exempt from reforms land that held cultivated mulberry trees. The significance of these acts was huge: in general, agricultural land was historically taken away from peasant owners, but at some point not only was land with mulberry trees not confiscated, it was the only land that could be passed on to family heirs.

All the steps of silk production were relegated to women: feeding and caring for the moths throughout their life cycle, collecting cocoons, reeling (unwinding) silk filaments, and spinning and dyeing thread. Women also wove the silk threads into fabric, although that was done at looms in special "workshops," rather than in individual homes. Apparently, some women even kept silkworm eggs warm by putting them in a fabric bag under their clothing. Women are considered to have made the improvements in both the loom and silk-reeling frames. Because historical sericulture was a home-based activity, women taught their daughters to master all the steps of production, thus continuing the activity across generations.

As with all moths and butterflies, *Bombyx mori* has a complete life cycle: egg, larva, pupa, and adult. The eggs are pinhead-size, so small that it would take 1,000 of them to tip the scales at 1 gram. A female silk moth can produce only about 500 eggs over a few days, so two females would be required to yield 1 gram of eggs. For the first step in production, about 1,000 eggs are placed onto large bamboo trays and covered by mulberry leaves. The larvae, once hatched, multiply their weight 10,000-fold in roughly a month if kept at the ideal temperature and conditions (Figure 1.4).

Read that again—a 10,000-fold weight gain in 30 days! The initial gram of eggs would now be 10 kilograms (about 22 pounds) of larvae, or caterpillars. Caterpillars can be considered the "teenagers" of the insect world. Just like human teenagers, insect teenagers have ravenous appetites. However, even human teenagers don't gain that much weight, thankfully. Parents would need more or larger refrigerators in their home. As the mulberry leaves are consumed, they are replaced with fresh

Figure 1.4 Silkworm (*Bombyx mori*) caterpillars on tray of white mulberry (*Morus alba*) leaves.
Source: Socrates 471/Shutterstock.com

leaves, and this process requires that the larvae on their tray be cared for almost continuously. Four trays of 1,000 eggs, even accounting for larval mortality, eventually will produce about 1 kilogram of silk. One kilogram? Silk is very light, so 1 kilogram is a *lot* of silk, about enough to make one skirt and blouse combination or about 60 neckties. Yes, one kilogram of silk *is* a lot of silk. But we're jumping ahead.

After chewing on mulberry leaves for nearly a month, each fully developed larva stops feeding and begins to spin a cocoon made of silk. Silk is secreted from the paired salivary glands. Just as you have salivary glands, so, too, do silkworms. Yours are important; a silkworm larva's are hyper-important. As an indication of their importance, consider how large a silkworm larva's salivary glands are. If the caterpillar's salivary glands were stretched to their full length, they would measure 10 times the length of the full-grown caterpillar and they would make up half of its body weight. Think back to your own salivary glands—half your body weight? Don't think so. Let's do the math. Assume your body mass is 100 kilograms. You have three primary salivary glands, each as a pair. In total, they have a mass of about 96 grams (about 3.4 ounces). Your salivary glands represent 0.096% of your body mass, versus 50% for a silkworm's salivary glands. Apparently, silkworm salivary glands are pretty important to them.

The moth's saliva is liquid while inside the moth's body, but it hardens once the filaments from the salivary glands are exposed to air, the filaments adhering to each other and forming a single strand. The process of spinning a cocoon takes several days, with the moth spinning at a rate of nearly 15 centimeters per minute. The cocoon is formed by a single strand of silk more than a kilometer in length. Let that sink in for a moment. Silkworm cocoons are made of a single strand, thin as a human hair and more than a kilometer long. That fact alone attests to silk's strength. Maybe no suspension bridges are made of silk, but it is a mighty strong natural fiber.

The completed silken cocoon contains a mature larva, which undergoes pupation as it transforms from a larva to an adult in the process of metamorphosis. The bamboo tray of larvae is now a bamboo tray filled with silkworm pupae (Figure 1.5). Pupation is often thought of as an inert, inactive part of an insect's life cycle—a pupa just sits there, doesn't move, doesn't feed, looks like it doesn't do anything at all. Far from it! Physiologically, the pupal stage is the most active part of the insect's life cycle, as it transforms from a caterpillar to an adult with wings. To do that, enzymes break down larval tissues and use the remaining building blocks to produce an adult moth, with no further external inputs, out of view and all within the spatial confines of the pupal case. This would be the equivalent of construction engineers hanging a large sheet over a three-story brick walk-up and, with no new construction materials brought in, a week later pulling off the covering sheet to reveal a shiny new glass-covered building complete with condominiums, new plumbing and wiring, and an elevator, ready to sell to upwardly mobile urban residents. Moth pupation as an analog to gentrification.

In nature, the pupal stage ends with the emergence of an adult from the pupal case formed inside the silk cocoon. For domestic silk moths in production regimens, that process is intercepted at the appropriate time to harvest silk from the cocoon, and

Figure 1.5 Silkworm cocoons on trays.
Source: Yargin/Shutterstock.com

the timing of the interruption is critical. Cocoons need to be mature enough to yield high-quality silk but not so far advanced that the quality may diminish. A new adult moth secretes enzymes as it prepares to emerge from its cocoon, and as it emerges it breaks the cocoon, thus breaking the continuous thread into many threads that are harder to weave. Also, the older the cocoon silk, the less readily it takes up dye. So, cocoons are treated with steam to kill the enclosed insect (only one insect develops per cocoon) and then further treated by immersion in hot water—or an empress' tea—to loosen the silk strand for harvesting (Figure 1.6). A loose end of the silk strand that makes up a cocoon is attached to a spool, along with the filaments of 5–8 other cocoons, and all are spun together to form a single, strong thread, suitable for transformation into fine—and valuable—silk garments.

SILK AS CURRENCY

During the Han Dynasty, silk assumed its own intrinsic value, much like precious metals such as gold or silver. Once silk was known throughout China and in external regions, it became a desired commodity. Lengths of fine silk were used to establish monetary values, and silk became a measure of currency throughout the strata of society and across the breadth of the empire. Silk became universally accepted as monetary currency for trading and paying debts, especially where coins were not used as currency, such as for trade outside of early China. Agricultural producers paid some or all of their taxes in silk, and soldiers in distant postings were paid in silk.

Figure 1.6 Collecting silken threads from silkworm cocoons.
Source: GG6369/Shutterstock.com

Silk was also the primary currency for payments made to keep peace with the nomads of the Mongolian Steppe. Paying for peace amounted to little more than extortion, but the Chinese were willing to make the payments to avoid attacks on their outlying cities. One fierce tribe, the Xiongnu, was the most feared and most demanding of the nomads. Their tribal leader, the *shanyu*, periodically "requested" meetings with the Chinese emperor. At those meetings, the *shanyu* collected his "gifts," which consisted of rice, wine and, most desired, silk. Silk was a symbol of power and status for the *shanyu*, and so it was highly desired as a tribute payment.

Maintaining peace and a trading partner was an unequal trade exercise, with valuable goods flowing to the Xiongnu and the only reciprocity some ill-defined period free from raids. The staggering cost of payments continually increased, but the cost was considered to be less than losses from a raiding tribe and so the emperor was willing to pay—up to a point. As with all extortion ploys, this one reached a tipping point, and the emperor was advised that paying the escalating cost of maintaining peace with the Xiongnu was no longer wise. One might say that the greed or status-seeking of the Xiongnu was like a later version of Aesop's fable of the goose and the golden egg, in this case a silken egg.

Regardless, by 119 BCE, the Chinese had taken control of the Xiyu province in the far west, which was a rich land for agricultural production. Through a series of economic controls and exertion of military might by the Han, the Xiongnu were transformed from marauders to become a part of the empire, though still receiving payment from the emperor. However, the entire system had changed, as the payments were made in exchange for homage paid to the Han emperor. Reciprocity by the Xiongnu at that point was their submission, for which the Han were willing to pay—what had begun as paying for peace had become paying for homage.

The system of trade, diplomacy, and paying for extortion in exchange for peace all happened because of silk—held by the emperor and desired by the tribe's *shanyu*. While keeping peace with the Xiongnu, the payments by the emperor reflected the size of the accompanying retinue visiting the emperor's court, as the *shanyu* distributed silk among those accompanying him. For a number of years, the emperor limited the number of the *shanyu*'s entourage to 200 courtiers. However, for the visit in the year 1 BCE, the Xiongnu *shanyu* requested that he be allowed 500 followers. Begrudgingly, the emperor relented, which resulted in payment of the unfathomable amount of 30,000 catties of silk (a cattie is roll of silk weighing half a kilogram, or just over 1 pound), the same number of fabric pieces, and 370 suits of clothing.

Let those amounts sink in for just a moment while we consider how that payment could have been made. Just the silk rolls alone amounted to 15,000 kilograms of finished silk. That is 15 metric tons of silk. Imagine the effort by the Han emperor and his empire to sustain that level of silk payment, initially for extortion, and only later for submission and homage. Consider the entire supply chain: growing mulberry trees, feeding the silkworms, harvesting the silk, weaving the cloth, then collecting the products from homes and workshops. From how many individual homes and workshops, and over what area, did the empire need to collect to make a payment of 30,000 rolls of silk plus the fabric and items of clothing? And what about the army of workers that would have been necessary to coordinate and make those collections, plus deliver that amount to the emperor's court? And then how did the Xiongnu get their "parting gifts" home?

SILK AND OTHER ANCIENT FABRICS

As stated previously, for the first few thousand years after production of silk began, its use for clothing was restricted to the emperor, his family, and selected nobility. Well, that's fine, but what did everyone else wear? Good question. To answer, we need a short detour into the history of textiles. What did commoners wear? Clothing made from linen, hemp, ramie, cotton, and wool. All five have ancient origins, each vying for "the oldest of fabrics" but with the exact dating of any of them uncertain. Let's take a closer look.

Wool, although seemingly obvious as an ancient source of fabric, may have been the most recent of the five to have been used as a textile. Domestication of sheep began around 10,000 years ago, but early sheep were more "hairy" than "woolly." Breeding is thought to have begun about 8,000 years ago, and only after many years of breeding was the shorn wool suitable for weaving.

Evidence of woven woolen fabrics appeared about 3,000–4,000 BCE in southwest Asia. The use of wool began because of its ease in spinning and because its fibers are bulky and retain air, thus providing both warmth from cold temperatures and insulation from extreme heat. In its early use, wool was collected either by hand-plucking or through the use of special bronze combs. Shearing of sheep did not begin until much later, in the Iron Age (roughly 1,500–2,000 years ago in Asia).

Cotton figured prominently in the history of textiles and clothing worn by all ranks of people. The plants that produce cotton are descended from perennial shrubs or small trees, which humans have domesticated and transformed into plants that are now grown as annual crops. These crops are part of a collection of about 50 species considered in a single genus (*Gossypium*), and four of them have been domesticated, two in the Americas, the other two in Africa and Asia.

Most cotton grown in modern times has been *Gossypium hirsutum*, one of the species from the Americas that has been moved around the world. The other South American species, *Gossypium barbadense*, is curiously called Egyptian cotton. Cultivation of the two species from the Americas began thousands of years later than cultivation of the African and Asian species.

The African species, *Gosssypium herbaceum*, is thought to have had its ancestral home in in northeastern Africa, southwest Asia, the Arabian Peninsula, and Syria, where it was already domesticated widely as a crop 3,000 years ago. Indeed, the origin of the word "cotton" is Arabic: "al qutn," later becoming "algodon" (Spanish), "coton" (Old French), and then "cotton" in English. One variety of this species is thought to have been the first cotton grown as an annual crop, and this species—not the one from South America—would be the correct one for humans to give the moniker Egyptian cotton.

An Asian species, *Gossypium arboreum*, originated approximately 8,000 years ago on the Indian subcontinent, in the Indus Valley, in what are today's India and Pakistan. The earliest archaeological evidence of cotton comes from a Neolithic site located in Pakistan, where the northern edge of the Indus River basin meets the southern edge of the Iranian Plateau. *Gossypium arboreum* is a subtropical and tropical plant (as are other *Gossypium*), and producing the cotton fruit, or boll, required long daylength. As a result, its cultivation was limited to southern Asia until after the first millennium BCE, when breeding changed the plants' reliance on long daylength to produce the fruit. The fibers formed in the cotton boll could be spun into thread and made into clothing.

Hemp (*Cannabis sativa*), another of the ancient textiles, was in use at least by 6,000 years ago, as shown by archaeological evidence from China. This annual plant has a strong, hollow stem that, with physical processing, yields long and strong fibers that can be separated for use as a textile. The coarse fibers are ideal for use in making string, rope, and burlap, and, because of its strength, it was used to make canvas, such as that used as sails for early ships. The word canvas comes from the Latin word for hemp: *Cannabis*.

The early Chinese used hemp for rope and paper. As a textile, it had limited use because it was coarse and did not bleach or dye well. Hemp was, however, combined with other fibers, such as flax. Modern processing has led to hemp being used as a specialized fabric resembling linen. An interesting use of the plant came from Japan where, from the middle of the 18th Century and lasting for 200 years, hemp fibers were woven to make mosquito netting, called kaya. Because of its ubiquity in everyday life, kaya made from hemp even featured as the background or subject of some

Japanese paintings. Contemporary renewed interest in hemp as a fiber crop has expanded beyond producing textiles and has led to its use for bioplastics and also as a recreational pharmaceutical.

Ramie (*Boehmeria nivea*) is a plant native to China and East Asia, whose use as a textile goes back to at least 5,000 years ago. It is a herbaceous perennial, classified in the nettle family. Like other nettles, the underside of ramie's leaves has numerous fine hairs, but the hairs do not sting as do those of its close relative, the stinging nettle. Another close relative used for fabric is the variety that is native to south Asia known as rhea, with no hairs on the leaves.

Ramie has had various names over the centuries, including China grass. Its stalks can grow to more than 2 meters tall and are harvested by hand several times per year. Harvesting the plant is simple; collecting the fiber is much less simple. The plant produces a bast fiber, meaning the fiber is derived from the phloem of the plant, not the inner part of the stem nor the bark itself. After the stalks are harvested, the bark and underlying fiber are separated from the inner stem by soaking in water. Once separated, the bark is scraped to yield fibers adhering to the inner bark, in a process called decortication. The fibers themselves are held together by sticky resins, which require chemical treatment to separate.

The fiber of ramie is white and lustrous, resembling silk. Spinning the fibers into thread is not simple because they are brittle and the resulting thread is hairy, rather than fine. Removing the sticky resins during harvest also takes away the fibers' cohesion. But, once ramie is woven as a textile, it is stronger and more durable than wool or cotton and has acquired colloquial names, such as grass linen or China linen. The inelasticity of ramie fabric means it does not lose its shape when wet. Its resistance to bacteria and mildew explain why it has been alleged to have been used as an outer shroud for wrapping mummies. However, the complex harvest process has limited the general acceptance of ramie, outside its uses in eastern Asia.

Linen has had a long history and broad usage worldwide, rivaling that of silk. The source of linen is the annual crop plant, flax, *Linum usitatissimum*. The word linen has its ancestral lineage in the Greek word "linòn," later becoming the Latin name of the genus, *Linum*, which ultimately became the currently used linen. Flax has been cultivated for at least 9,000 years, though some sources claim its cultivation occurred even earlier. Flax fibers are longer, stronger, and smoother than those of cotton and finer than those of hemp or wool. With its long history, flax that was woven into linen made the textile a contemporary of silk, at least in China.

Flax and linen have inserted themselves into common usage beyond textiles. Because flax fibers are long and strong, plant fibers were processed to make thread, which was used for weaving into fabric. The Latin word *Linum*, gave rise not only to the name of the fabric, but the fine threads used for linen also were the origin of the English word, "line." A linen thread, or line, was used for measuring and it defined straightness. Linseed oil also gets its name from the Latin *Linum*, as does the word "lining," whose usage had its origin as the inner layer, or lining, of many garments.

Woven linen was used for funerary shrouds, and linen-wrapped Egyptian mummies, found with the linen in apparently perfect condition thousands of years after entombment, attest to its longevity and strength. Linen textiles are still prized for their "cool" feeling and lack of lint, giving the surface a clean look. In modern times, flax has been grown more as an oil seed crop than as a fiber crop harvested for textiles.

Linen may rival silk in its age and fineness as a fabric, but the product of silkworms, made from moth spit, has had a far more storied life of its own. Silk was the desired and valuable product that stood the test of time. Ancient in origin, vibrant still today. Used as currency, whether to buy wares or protection, silk provided the basis for extended transcontinental trade, created great empires and formed the threads that connected cultures and ideas. What began as sticky threads from blind, flightless moths, connected humanity. The story of silk is the story of humans.

CHAPTER 2

༺

The Silk Roads

Zero degrees longitude, the prime meridian, is where east meets west. No actual line marks the limits of the two hemispheres; instead, an imaginary line passes through the Royal Observatory in Greenwich, England, from where we get the concept of Greenwich Mean Time. The prime meridian is the heart of the Western world and all things Anglo. Much of Western history is based on the prime meridian, but it isn't the heart of all human history. Many of us learned when we were young that the heart of human history, the "cradle of civilization," was farther east. On that, we all would likely agree. From there, opinions will vary widely, depending on interests and perspectives. The Western bias that many of us learned certainly tends to follow history as far east as Egypt and the eastern end of the Mediterranean, particularly the 2,000-year-old history associated with Christianity, and the Greek and Roman Empires. Usually, scant attention is paid to the rich history of the Chinese dynasties. And the area between the Mediterranean and China? And its complex and uber-rich history? Many of us know more about our solar system than we do of the history of Central Asia.

We are not offering here the definitive history of Central Asia, one that would stand up to scrutiny from professional historians. Our academic field is entomology, not ancient history. However, that vast geographical span from the Mediterranean to the Pacific Ocean, with its rich history, encompassing economics, science, religion, and war, is of great interest, and not just to entomologists. The reason—the thread that connects East and West—is an insect: the silkworm, *Bombyx mori*. We chose the silkworm as one of the five insects for this book because of its influence on human history. The product of silkworms is, of course, silk. The history of silk is the history of humankind, and the silken thread that ties together much of history and vast geography was woven along the Silk Roads.

The landscape we paint of the Silk Roads spans from eastern China to Constantinople, the modern-day city of Istanbul. Much of the history in this chapter occurs in the countries whose names have the suffix "-stan": Afghanistan, Kazakhstan,

The Silken Thread. Robert N. Wiedenmann and J. Ray Fisher, Oxford University Press. © Oxford University Press 2021.
DOI: 10.1093/oso/9780197555583.003.0002

Kyrgyzstan, Tajikistan, Turkmenistan, and Uzbekistan. That part of the world is dominated by mountains and desert. The formidable high-elevation Taklamakan Desert occupies most of the Tarim Basin (Figure 2.1). Although it is in the vicinity of the oldest civilizations, the Tarim Basin was, ironically, one of last places in Asia to be inhabited. That area of Asia is where great mountain ranges converge: to the north, the Tien Shan, even farther north, the Altai. To the west lie the Pamirs, Hindu Kush, and Karakorams; to the south, the Kunlun Shan then the Tibetan Plateau and the "roof of the world," defined by the Himalayas, which were created when the Indo-Australian tectonic plate crashed into the Eurasian tectonic plate, pushing up that great mountain range, which is still rising by 2 centimeters a year.

Sprinkled amid and beyond this geographic and geological enigma are cities and towns, outposts with exotic names and equally exotic histories: Bactra (namesake of the Bactrian camel), Bagram, Bukhara, Dunhuang, Kandahar, Merv, and Samarkand. Ancient cities all, several were founded by Alexander the Great as his conquests took him eastward in the 4th Century BCE. Alexander re-named existing cities after himself, three of them in modern-day Afghanistan: Alexandria in Aria (Herat), Alexandria in Arachosia (Kandahar), and Alexandria of the Caucasus (Bagram). Herat was nearly obliterated by raiding Mongols, and Merv was so devastated by them that the city exists today only as a desolated ruin. Great historical events and figures are

Figure 2.1 Satellite image of the Tarim Basin, Xinjiang, China; with the Tien Shan Range to the north (top), the Kunlun Shan Mountains to the south, and the Pamir Mountains to the west. Note the enormous dust storm in the eastern Taklamakan Desert, which occupies most of the basin.
Source: Courtesy NASA

enmeshed in these cities, all with stories to tell. Those cities and many others may seem to be sprinkled across Asia, but their locations are not random. They were outposts and trade centers along the web of routes collectively known as the Silk Roads (Figure 2.2), routes that crisscrossed Asia and were traversed by traders for more than 1,500 years.

NOMADS, HORSES, AND SILK

The Silk Roads were a web of routes crossing the Asian continent, making connections across space and through time. They were a magical, mystical, metaphysical realm of international trade, ancient history, and diverse people. The Silk Roads are famed today for their antiquity, but they were not the first routes of trade to exist in Asia. At least 7,000 years ago, Eurasian pastoralists, the communities of people who domesticated animals, established an extensive system of trading routes, eventually spanning more than 9,000 kilometers from the Black Sea to East Asia. These so-called Steppe Roads began as simple trading routes formed when nomads began to settle into semi-sedentary communities. Early efforts to domesticate plants and animals generated surpluses, which led to trading what one community had for what another produced—an economic system of exchange that was new to the world.

Warming temperatures in northern and central Asia about 5,000 BCE expanded the extent of the grasslands and allowed new plants, such as wheat and barley, to be cultivated as crops. The grasslands also fed animals that had been domesticated, the most notable being Bactrian camels and Mongolian horses. As some people settled into an agrarian lifestyle, others retained nomadic movement in search of pasture for their domesticated animals. The two lifestyles developed simultaneously—agrarians produced food and nomads served as traders of that food. Animals domesticated by the nomads about 4,000 BCE enabled trade and communication across the Steppe Roads. Bactrian camels, domesticated in the region of Bactra (today's Central Iran), could carry heavy loads over long distances—even laden, they could travel at a speed of 3–4 kilometers per hour for up to 10 hours per day. They later became the iconic animal symbolizing the caravans transporting trade goods on the Silk Roads.

Even more so than camels and before them, Mongolian horses—once domesticated and selectively bred—enabled the importance of the early Steppe Road trade routes. The horses were themselves prized and became a key trade item, a commodity with multiple uses for the Kurgan Horse Culture, one of the "horse cultures" prominent in 4,000–5,000 BCE. In addition to their value as transport, horses gave the Kurgans important protein: meat and milk. Horses were also traded for food or other precious commodities. The Kurgans, as extensive travelers whose influence covered large areas of Eurasia, traded their horses for gold, silver, and gems, thus establishing horses as currency on an equal standing with precious metals.

The Silk Roads were made possible only because of the trade network of the mostly overlooked Steppe Roads. The Steppe Roads gained their importance because of the convergence of a warming climate, which expanded grasslands as a grazing habitat, and the nomadic Kurgan people, who domesticated animals and traveled in search

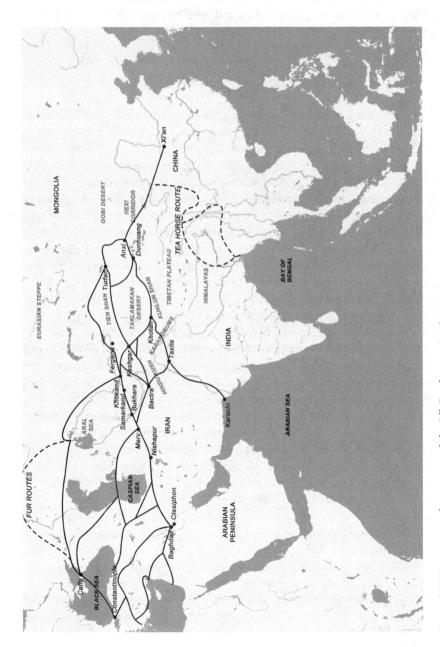

Figure 2.2 Major routes that comprised the Silk Roads spanning Asia.
Source: Sarah Silva, MapShop.com

of forage for their animals. As the nomads traveled, they interacted with others and exchanged surplus goods with those they met. The domestication and commodification of horses established links of commerce. All of this occurred in the steppe grasslands of Central Asia, and all of it was necessary for the Silk Roads to become the nexus, the link of events and activities, that established the trajectory of humans over thousands of years. Overlaid by climate, the critical junctures shaped history, economics, emergence of power, and development of food systems. All this because of the products from moth saliva: from nomads and horses to silk.

THE ROYAL ROADS

Other nomads appeared a few centuries later than the Kurgans: the Scythians (Table 2.1). This group of pastoralists was thought to have originated about 3,000 BCE in what today is known as Iran. Unlike the period of the Kurgans, the cultural prominence of the Scythians lasted for thousands of years, extending until the Middle Ages. At their peak, the Scythians controlled the steppe from China to Eastern Europe. Theirs was a complex culture, and they were far more than traveling pastoralists seeking forage for their animals. They appreciated artwork and precious metals, but they were also a much-feared, warring people known for mounted attacks with bows and arrows, predating by 1,000 years the methods iconified by the Mongols. But the Scythians also made the Silk Roads possible by improving trade, making it faster and more efficient, and by trading over an area that reached from China to Egypt. The Steppe Roads were the necessary precursor to Silk Roads. Necessary, but not sufficient.

To the southwest of the steppe lay the land that became the empires that defined the Persian Dynasty. The land of the Persis, the geographic extent of the empire, was itself spanned by major routes traversed by traders and armies. Beginning in the middle of the 6th Century BCE, at the onset of the Achaemenid Empire (Table 2.1), the route knitted together that empire, extending 2,500 kilometers from Greece, Anatolia, and the coast of Asia Minor eastward to Babylon (in modern Iraq) and Persepolis (in modern Iran). Baghdad, founded about the time Babylon was abandoned, was at the crossroads along the major routes of trade, both regionally as well as externally—throughout Asia, East Africa, the Mediterranean, and Western Europe. Exotic items traded in and beyond Baghdad included ivory, gems, and soap; many of the items arriving at Baghdad were exported further afield. Other major commodities that were traded included the crops that were produced where agriculture is said to have begun—that fertile strip of land lying between the Tigris and Euphrates Rivers. Even today, harvesting grain crops in that land includes "sorting wheat from the chaff" by tossing the grain in the air and letting the wind carry off the lightweight chaff.[1]

Revenue from trade allowed the Persian rulers to create lavish palaces and other prominent structures, buildings that incorporated luxurious materials such as gold, lapis, ebony, cedar, turquoise, and ivory, all acquired from afar on trade routes that connected eastern Asia with the Mediterranean. The same trade routes

Table 2.1. DYNASTIES AND EMPIRES, THEIR PRIMARY PERIODS OF ACTIVITY, AND THE REGIONS (MODERN-DAY) THEY OCCUPIED DURING THE TIME FRAME OF THE SILK ROADS.

Dynasty or Empire	Active Period	Regions or Territories (peak)
Central and Western Asia to Mediterranean		
Early Nomadic		
Scythians	7th Century BCE–3rd Century CE	north of Black and Caspian Seas; Tian Shen to Hindu Kush
Persian Dynasty		
Achaemenid Empire (First Persian)	550 BCE–330 BCE	northern Greece, Turkey, Black Sea to Aral Sea; south to Arabian Sea; northern Egypt to Arabian Peninsula
Parthian Empire (Middle Persian)	247 BCE–224 CE	eastern Turkey to eastern Iran; south to Arabian Sea; northeastern Arabian Peninsula
Sasanian Empire (Neo-Persian)	224 CE–651 CE	western Turkey to eastern Iran; Aral Sea south to Arabian Sea; Egypt, Arabian Peninsula (partial)
Other Empires		
Seleucid Empire	312 BCE–63 BCE	southern Turkey, Levant, Iran, Iraq, Kuwait, Afghanistan, Turkmenistan
Byzantine Empire	286 CE–1461 CE	southern Spain, North Africa, Egypt, northern Mediterranean, Turkey
Arabian Empire	632 CE–1258 CE	Spain, Portugal, North Africa, Egypt, Levant, Syria, Armenia, Iran, Iraq, Arabian Peninsula, Afghanistan, India
Mongol Empire	1206 CE–1368 CE	Mediterranean to Korea
Eastern Asia		
Non-Dynastic Tribe		
Xiongnu[1]	209 BCE–216 CE	Mongolia, northern China, Kazakhstan, Kyrgyzstan
Dynasties[2]		
Early Han (Western)	206 BCE–9 CE	China, Mongolia, Vietnam, Korea; west to Tarim Basin
Late Han (Eastern)	25 CE–220 CE	China, Mongolia, Vietnam, Korea; west to Tarim Basin
Tang	618 CE–907 CE	China, eastern Asia

[1] The Xiongnu were a long-lived nomadic tribe, not an empire.
[2] Only the primary Chinese dynasties associated with the Silk Roads are listed.

were also used to support Persian military outposts, situated to defend the empire. Raids came from the steppe to the north, in the form of nomads who were equal part fierce marauders and important traders of cattle and horses. The established road network also allowed invaders easy access. When Alexander the Great set his sights on the heart of the Achaemenid Empire, the Royal Road allowed his army to move quickly and capture city after city, creating his own short-lived empire. Interestingly, Alexander is remembered for his 13-year "empire," but its successor, the Seleucid Empire (Table 2.1), with a 250-year run, is seldom mentioned. More than 1,500 years later, both the Royal Road and the Silk Road carried Marco Polo eastward toward his meeting with Kublai Khan (discussed in Chapter 14). This area of Southwest Asia and these routes would feature prominently in the history of silk for centuries to come.

THE SILK ROADS—HEXI CORRIDOR AND TARIM BASIN

The term Silk Roads was first expressed (in German, *Die Seidenstrassen*) in 1877 by the German geographer Ferdinand von Richtofen. He was a renowned geographer and geologist, whose methods of geography and major publication on China are still relevant. Despite his academic fame and accomplishments, this was all overshadowed by that of his non-geographer nephew Manfred von Richtofen, whose legends were created in the skies over Western Europe in the First World War. It's worth noting that the term is plural (Silk Roads), as Richtofen noted, not singular (the Silk Road) as is often used. This is because Richtofen recognized the Silk Roads were a network, a web, not a single line.

The Silk Roads, considering their impact on countless aspects of history,[2] never had a formal opening or date of initiation. The use of these particular routes for trade, intersecting somewhat with the Steppe Routes, can be traced back to the Han Dynasty of China, which began its 400-year run in 206 BCE (Table 2.1). The Han had expanded their sphere of influence westward, where outpost cities fell under periodic attack from Mongol raiders, most notably the Xiongnu (Table 2.1). As "neighbors," the Xiongnu were a tribe best kept at arm's length. They were valued as trading partners, providing cattle and, importantly, horses, to the Han rulers, but they were equally reviled for their raids on the outposts (see also Chapter 1). The Xiongnu were also paid in silk not to attack the Han Empire's outposts, requiring unfathomable numbers of silkworms and an equally immense amount of silk from their cocoons. Ultimately, the conflict between the Xiongnu and the Han Emperor devolved into a decade-long fight, ending in 119 BCE, at which point trade extended beyond the western periphery of the Han Empire.

The network of trading routes called the Silk Roads ran along a main axis from the eastern Chinese city of Xi'an to its eventual western terminus in Constantinople (today, Istanbul), a distance of 7,000 kilometers (Figure 2.2). Modern-day countries that were major parts of the network included China, Tibet, the aforementioned "-stans" (Pakistan, Uzbekistan, Afghanistan, etc.) to Iran, Iraq, and Turkey. Extensions along the axis included Japan and Korea to the east and Greece and Italy to the west. Perpendicular extensions converted the road to a network of roads, including spurs to Southeast Asia, India, North Africa, Russia, and northern Europe, and maritime routes that both extended the network and bound it all together.

The Silk Roads—and here the plural matters—were a web of paths, some constant and continuous, others shifting, expanding, branching, or ending, responsive to the environment of that moment for reasons as varied as wars and climate. From the eastern beginning point in Xi'an, the road began as a singular route, passing westward through the 2,000 kilometer Hexi Corridor to Anxi, with the route constrained to the south by northeastern reach of the Kunlun Shan Mountains and to the north by the great Gobi Desert (Figure 2.3). The Hexi Corridor provided travelers some protection in the form of the Great Wall, whose length had been extended to its northwestern terminus at the Jade Gate, near Dunhuang (Figure 2.3). In exchange for protection, the empire's garrisons along the Hexi Corridor taxed the travelers. A 2,000-year-old toll road! Too bad there were no radar guns to catch speeding camels.

West of Anxi is the arid Tarim Basin, bounded all around by mountains (Figure 2.1). More than 900,000 square kilometers in size, the basin holds in its center the inhospitable Taklamakan Desert, which covers more than one-third of it (Figure 2.4). The Taklamakan has an annual rainfall of about 10 centimeters, limiting any plant and animal life to the immediate vicinity of the Hotan River or the ephemeral Tarim. All drainage from the surrounding mountains collects in the basin, and the few internal rivers largely dry up in the desert. Taklamakan sand dunes reach 150 meters in height and, coupled with strong winds, can create dust storms that blind travelers in caravans, with the clouds of sand dust reaching 4,000 meters in altitude.

Silk Road routes offered travelers choices, each with its own perils. The Northern Route passed to the north of the Tien Shan Mountains, avoiding the desert, and a spur from that route tracked to the north of the Caspian Sea, ending at the Black Sea; this route followed stretches of the earlier Steppe Roads. However, it was a long route that missed the major trading centers. Most travelers chose one of the more central tracks through the Tarim Basin. The Southern Route, which was the main route used early in the Silk Roads' history, passed along the south edge of the Taklamakan and along the foothills of the Kunlun Shan. This route was also known as the Jade Road for its role in the importation of Hotan jade into China. The Middle Route gained prominence during the Han Dynasty as a cutoff that reduced traveling distance and time, but the shorter distance also bypassed several important oasis stops. It eventually fell into disuse when the marshes in the eastern region of the Tarim Basin dried and ultimately disappeared, greatly limiting travelers' access to water. The Northern Route passed between the Tien Shan and the Taklamakan and was used most often after the Tarim River changed its course, partly because of the numerous oases and towns that provided respite and the supplies for replenishing.

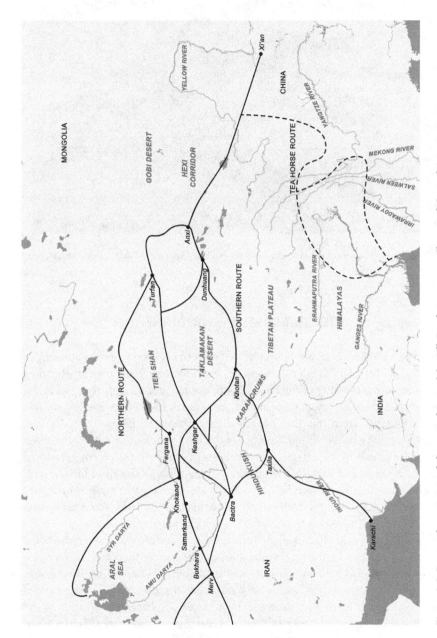

Figure 2.3 Routes that comprised the central core of the Silk Roads in western China and Central Asia.
Source: Sarah Silva, MapShop.com

Figure 2.4 Flaming Mountains in the Tarim Basin and the edge of the Taklamakan Desert, near Turpan, Xinjiang, China.
Source: Munzir Rosdi/Shutterstock.com

THE SILK ROADS—SOGDIANA

The Northern and Southern Routes met at the western edge of the desert at Kashgar, but westbound travelers' challenges were not yet over. Merchant travelers heading west still faced the crossing of the Hindu Kush, the mountain range that sits at the convergence of the Himalayas, Karakorams, and Pamirs and divides the Indus River valley to the south from the Amu Darya valley to the north (Figure 2.5). The history and geography of the Hindu Kush alone could fill volumes, and it is of modern relevance for its role in the defeat of several western armies in the recent decades of fighting in Afghanistan. Traveling west after passing through the Hindu Kush, traders reached the crossroads town of Taxila. Taxila was a prominent trade center until the 5th Century CE, adding a north-south route through the Indus River valley that connected cities in India with the Arabian Sea and the Maritime Routes.

Despite the physical challenges of traversing the routes, trade flourished. The period of the 4th–8th Centuries saw the apex of trade along the Roads, with much of the trade both enabled and dominated by the Sogdians (Table 2.1). Their home, Sogdiana, was a region located west of the Tarim Basin and along the core routes of the Silk Road, but its history predates the trade routes. Sogdiana was already a prominent area, with extensive tracts of productive, irrigated agricultural lands, when it was conquered by the Achaemenids about 539 BCE; it was conquered again by Alexander the Great in 332 BCE. After the Silk Roads came into existence, Sogdiana was a respite for travelers, with major oasis communities. The Sogdians helped connect ideas and transfer goods among distant regional powers, becoming wealthy

Figure 2.5 Snow-capped peaks of the Hindu Kush Mountains. Some routes of the Silk Roads traversed passes through the Hindu Kush.
Source: Valerii_M/Shutterstock.com

as both agriculturists and traders, and as a culture that made large-scale trading possible.

The Sogdians were multilingual and accepted others freely, so they could communicate and do business with a broad range of foreign traders. Located at the crossroads of several routes of the Silk Roads, the Sogdians participated in the trade of silks as well as other valuable goods, such as furs from the north, gems from the south, and horses from the adjacent, rich Ferghana Valley to the east. The Sogdians practiced and accepted a variety of religions, reflecting their open nature, multilingual ability, and location amid multiple cultures. The wealth that accumulated in Sogdiana enabled the development of large, permanent cities, such as Samarkand and Bukhara, which were fortified against the nomads of the steppe. Although open and peaceful, the Sogdians also engaged in military campaigns as needed to keep marauders at bay. For more than 800 years, the Sogdians defined the trade along the Silk Roads, until the arrival of the Arab Empire's armies in the 8th Century.

The Silk Roads formed a web of connecting paths, and not every traveling caravan took the same path, especially with the rise and fall of various empires. At various times, traders sought routes to avoid enemies or regions known to be the territories of thieves. For example, sometimes traders from the Roman Empire going eastward to Central Asia bypassed the fairly direct route through the territory of Rome's wartime enemies, the people of the Middle Persian Empire, the Parthians (Table 2.1). Instead, the caravans took the much longer routes that passed north and across the Caucasus Mountains. In other instances, the traders were willing to pay the Parthians as middlemen, for safe passage, protection from roving bands of thieves, and staying in Parthian cities. Although at times bitter enemies, at other times the Romans and

Parthians tolerated each other—Romans had money and were willing to pay, and the Parthians were willing to relax their animosity and accept Roman money, benefiting from the revenue. The practice of using silk to pay for safety or protection dates at least to the Han Dynasty and Xiongnu (Chapter 1), but it also occurred with payments made to Mongol leaders.

How did we get into all of this? Oh, right—the silkworm. Where does the silkworm fit into all of this? The silkworm's product, silk, was the impetus for the Silk Roads, but the diversity and value of the list of goods traded exceeded the original precious product. The Silk Roads were a network of commerce. The very existence of the Silk Roads illustrates how important commerce was to the people of the lands across Asia and also shows the value of products created from silk harvested from the silkworm. The journey on the Silk Roads was formidable. It was not just the long distances and marauding local people, but the travel required threading the needle of challenging arid and cold deserts and numerous high mountain ranges that would have deterred all but the most intrepid. Yet, the travel did occur and the roads were expanded over time. Where the routes of the network converged, trade centers and markets developed. Those trade centers attracted more people and, as the number of their residents grew, became the cities strewn across Eurasia.

CARAVANSERAIS

Between the trading towns were a series of caravanserais, rest houses that provided an important shelter to stop at on the route. Each caravanserai consisted of an inn for the merchants and an enclosed courtyard for the animals and their loads. Some structures were simple, with few rooms, others were extensive with space for many travelers (Figure 2.6). These inns were located within a day's travel from each other or a trading town, which allowed the short distances to be guarded by locals to ensure the safety of the travelers and their goods. With the development of trade routes and the increased magnitude and value of trade, caravanserais became more crucial and their number increased. Sales of thousands of horses annually provided an influx of money that was used to improve infrastructure associated with the Silk Roads and to add more rest houses. Caravanserais provided important places for information exchange—about the route ahead, the demand for certain items, customs necessary for trading with people of different backgrounds. The stopovers also exposed the merchants to different people, languages, clothing styles, inventions, foods, and cultures.

The entire one-way, east-to-west trip on the Silk Road, from Xi'an to Constantinople, for example, would have taken six months. Silk, as a secondary product of the silkworm's salivary glands, did not spoil even with a long transportation time. One curiosity about the Silk Roads is that particular goods may have been transported along the entire length of the route, but by a series of individual caravans. Each caravan traveled only short segments of the total distance, moving the goods between key trading towns. The trading network was porous, and some goods were traded with locals along the route. On arrival in one of the towns, where the caravans paused to

Figure 2.6 Caravanserai in the desert, along a Silk Road's route.
Source: Uwe Seidner/Shutterstock.com

rest, traders unloaded and sold or traded their precious commodities. Unsold goods were loaded, along with new local commodities, onto a new caravan made up of new merchants who then continued along the route. The original traders, having rested and traded their products, traveled with their new goods back to the town from which they originated and sold the new goods on arrival there.

SAMARKAND AND BUKHARA

The ancient city of Samarkand, in today's Uzbekistan, was well known in the past for its greatness. Due to its strategic location, Samarkand was conquered multiple times, by Alexander (329 BCE), Genghis Khan (1220 CE), and Timur (1370 CE) (hereafter, dates are assumed to be "Common Era" unless specified otherwise). Timur, also known as Tamerlane, made Samarkand his capital and turned the city into the most important cultural and economic center in Central Asia. Even today, its architectural beauty may be unrivaled in the Islamic world (Figure 2.7). Samarkand is also home to the Ulugh Beg Observatory, one of Asia's oldest observatories, initiated in 1424 (Figure 2.8). Ulugh Beg, who was the great-grandson of Timur, was a mathematical genius who could speak five languages and whose observations with the observatory's sextant allowed him to catalogue more than a thousand stars—with no telescope. He also calculated the duration of a year with astonishing accuracy, measuring it to within 25 seconds of the best measures made today with modern instruments. Ulugh Beg's skills in mathematics and astronomy dwarfed his non-scientific skills. As a ruler, his reign lasted only two years, ending with his murder by his oldest son. As a

Figure 2.7 Geometrical architecture and mosaic art, Samarkand, Uzbekistan.
Source: Hussain Warraich/Shutterstock.com

Figure 2.8 Ulugh Beg Observatory, Samarkand, Uzbekistan.
Source: Moehring/Shutterstock.com

husband, he had only 13 wives, a paltry number compared to the number of stars he catalogued. Maybe he did not have enough silk.

The oasis city of Bukhara, west of Samarkand and also in Uzbekistan, was founded at least as early as the 3rd Century BCE. Bukhara had covered bazaars that were a major center for trade and craftsmanship. Because of its location, Bukhara was of strategic interest and so was captured by multiple empires in numerous raids, including seizures by Genghis Khan and Timur. During Bukhara's prime, in the time between its founding and seizure by Timur, the city had a famed history. Bukhara produced great poets and became a key intellectual and cultural center, said to even surpass Baghdad as the leading cultural center in the world of Islam. At the beginning of the second millennium, Avicenna, the Persian intellect and polymath, of the sort that arises very infrequently, was born and raised in Bukhara. Avicenna, or Ibn Sina, is said to have memorized the Quran by age 10 and declared himself a physician by age 16. His knowledge and practice of medicine became legendary when his treatment is said to have cured the Emir of Bukhara, the ruler over the local territory.

Ibn Sina was indeed the polymath of the time. He has been considered to be the father of modern medicine, and his Canon of Medicine was the standard medical text for more than 700 years. In his spare time, Ibn Sina was a philosopher, writing on logic and ethics, as well as a being a skilled mathematician and astronomer. As a testament to his broad knowledge and influence, Avicenna is immortalized by botanists, who named a genus of mangrove trees *Avicennia*. There is no record of him contributing to entomological knowledge, but even polymaths cannot do everything. We'll cut him some slack. Think of him as an 11th-Century Isaac Newton; better, think of Newton as a 17th-Century Ibn Sina.

THE EMIR, THE ARK, AND THE PIT

Bukhara is also famed in entomological circles for the city's massive fortress, known as the Ark (Figure 2.9). Contained within the Ark was a prison and, deep within the prison, was the infamous "Pit," the ultimate in torture chambers. The Pit was a hole, 4 meters deep, covered by a metal grate and with the only access in or out by a lowered rope. The Pit was where the emir, Nazrullah Khan, ordered insect-inflicted torture on enemies, whose transgressions were real or imagined.

Nazrullah's torture was exposing prisoners to insects, such as lice and "kissing bugs," which caused painful bites, plus scorpions and rats thrown in for good measure. Bad measure would be more like it. As their name suggests, human lice (detailed in Chapter 8) infest humans, causing itching, which can be painful in the case of infestations by large numbers. Kissing bugs are a group of assassin bugs organized into one subfamily (Triatominae) that feed on vertebrate blood. They get their common name from a habit of some species, which puncture the skin near a person's mouth. Although the triatomine bugs of Central Asia do not carry the pathogenic protozoa (*Trypanosoma*) that causes Chagas disease in American tropics, they can become a problem when present in large numbers because of their blood-feeding behavior. Even without the added torture from creatures such as scorpions and rats,

Figure 2.9 Ancient Muslim architectural complex, Ark fortress, Bukhara, Uzbekistan; a UNESCO World Heritage Site.
Source: Zufar/Shutterstock.com

the extreme discomfort of being infested by lice and triatomine bugs would cause prisoners to lose appetite and, especially, lose sleep.

Bukhara's emir ruled the kingdom of Bukhara with extreme brutality, earning the nickname, "The Butcher." In December 1838, British emissary Colonel Charles Stoddart, who was tasked with delivering a letter of peace to the emir, made a considerable diplomatic faux pas: first, following British military protocol, he entered the Ark fortress on horseback instead of on foot; second, he saluted the emir from horseback instead of dismounting and bowing before him; and, third he arrived without the requisite gift for the emir. These three gaffes, plus a few more, were considered unforgivable transgressions. Stoddart was thrown into the prison and then into the Pit, where he was exposed continuously to the mental and physical torture of lice, bugs, rodents, and scorpions.

At this point, stories vary a bit regarding the time and sequence of imprisonment and Stoddart's forced conversion to Islam. Well, he had a choice—convert to Islam or be beheaded. A letter from Stoddart made it to the outside world where his plea for help was passed on to Captain Arthur Conolly, who was ordered to travel to Bukhara to try to win Stoddart's release. Conolly found a gaunt, emaciated Stoddart, weakened by his exposure to the Pit's torture. Further insults were inadvertently made by Conolly and the enraged emir ordered Conolly to join Stoddart in the Pit.

Accounts tell of "flesh gnawed from their bones." Perhaps. If so, do not blame the insects—neither lice nor triatomine bugs actually "gnaw flesh." Maybe the rats did. Or perhaps the description is just hyperbole. After a year (or more, depending on the tale), the insect-tortured emissaries were made to dig their own graves, after which the emir ordered the two to kneel before their graves, where their executioner stood with his sword. In a final act of defiance, Stoddart shouted out that the emir was a tyrant, resulting in a swiftly falling sword. Conolly, a devout Christian, was given the opportunity to convert to Islam. His refusal was swiftly dealt with. Two things to note about the treatment imposed on the two Britons: (1) both Stoddart and Conolly traveled in peace—what would their treatment have been if they had been bringing articles of war? And (2) their beheading may have been the only near-humane treatment ever ordered by the emir.

SILK, SPICES, RELIGION

Although transporting the product from silkworm moths may have been the impetus for developing the Silk Roads, other products soon filled the routes in caravans moving in different directions. From Mongolia, Siberia, and Central Asia came animals and animal products, such as cattle and horses and products associated with them, furs and walrus tusks, amber, copper, and apples; from India and southern China came precious stones, spices, herbal medicines, and cotton textiles; the Middle East and Mediterranean areas contributed food (dates, nuts, fruit, olive oil), jewelry, art, gems, and textiles. New trade products arose during the Tang Dynasty, beginning in 618 (Table 2.1). This was the age of poetry and art; poetry was not traded as a commodity, but art was. Tricolored glazed pottery, woodblock prints, and paintings, especially those made on silk, marked the period of the Tang Dynasty.

The expansion of the Silk Roads (Chapter 3) to include a web of maritime routes saw the arrival of spices, originating in the Spice Islands of Indonesia and the Pacific. Traders brought from the West three innovations: stronger horses, wheeled transportation (wagons, carts), and metallurgy. China responded by adding bamboo, mirrors, oranges, and ginger to the array of products traded. Gunpowder made its way west from China, as did paper, which was invented during the Han Dynasty and was moved to the Islamic world in the 8th Century. Still, silk—moth spit—was king, and its value greatly exceeded that of the other products.

One enduring impact of the Silk Roads may have been as routes of religion—they were crucial in the development and distribution of religions and their tenets. Every major religion was transported, with its ideas, along the routes, reaching all pockets of humanity throughout Eurasia. Travelers were exposed to and absorbed the different cultures formed around religions and carried the new ideas home with them. Buddhism was transported to the west and south with goods from China along the Roads. In the opposite direction, Christianity and Islam were passed along from the Eastern Mediterranean to south Asia via travelers on the roads, and to southeast Asia as a result of the monsoons carrying sailors east. Many of these religions were practiced simultaneously and with acceptance by the Sogdians, where the religions also adopted norms of others. Caravanserais were a crucial component, serving as loci for travelers to be exposed to, learn about, and benefit from other religions and belief systems. Silk to spices to religion.

THE DECLINE BEGINS

Trade, and the vitality of the Silk Road network, waxed and waned through time, and the routes branched, expanded, and took on new extensions at different times and places. Eventually the core of the Silk Roads declined, as many pacts, treaties, and empires do. Beginning in 119 BCE, the Silk Roads expanded westward with growth of the Han Dynasty (Table 2.1). The Han Dynasty is considered an age of prosperity and scientific advances, enabled by trade of goods and ideas, until its collapse in 220, when travel along the core routes was disrupted. Other empires filled the void. The Sasanians of western Asia (Table 2.1), the last of the Persian empires, had come into power in 224 with the fall of their predecessors, the Parthians, and the Sasanians remained the regional Persian power until 651.

Early in the 7th Century, Muhammad declared himself prophet and the Islamic religion was founded. Muhammad's followers raided merchant caravans in Arabia and their raids eventually led to the conquest of Arabia. From that base, the Islamic influence grew, especially after Muhammad's death in 632. The armies of the Arabian Empire (Table 2.1) were then led by Muhammad's closest companions, Islamic leaders known as caliphs. The series of caliphs continued Muhammad's territorial expansion, conquering the Byzantine Empire all the way west to the Iberian Peninsula and, later, the territories of the Sasanian Empire. That wave of territorial expansion by the Arabic Empire led to changes in people, religions, and powers along the Silk Roads.

Other stresses to the Silk Roads came from the An Lushan Rebellion (also called the An Shi Rebellion) in China, which occurred from 755 to 763. General An Lushan initiated the rebellion against the Tang Dynasty when he declared himself emperor in North China, establishing his own short-lived dynasty. The fighting in the rebellion and subsequent disruption to the order in China caused millions of deaths, many from mass starvation, and population shifts as people fled the destruction. As armies guarding the Western Regions were recalled to fight Lushan's forces, the Tang lost control of those areas for several years, particularly the Tarim Basin, the crucial bottleneck for travel on the Silk Roads.

The An Lushan Rebellion and its effects seriously disrupted commerce overall and was felt all along the Silk Roads. Although the rebellion lasted for only eight years, rebuilding was a long process. Also, subsequent wars and uprisings caused a migration of a large number of people from northern China to south China. The loss of territory, and of people through wars, starvation, and emigration, reduced the economic power and tax base during the recovery. Certainly the An Lushan Rebellion was a major blow to the Silk Roads, but it was only the start of a long spiral of decline.

Over the next few centuries, the expansion of Islam dominated the territories encompassing the Silk Roads, but they remained, with the Islamic caliphate extending to the Indus River. A fight with the Chinese at the Battle of Talas led to the defeat of the Tang, but also the transfer of knowledge to the Arab world of how to make paper, which would later be crucial for development of the printing press. The Islamic empire broke into smaller kingdoms in the 10th Century, allowing local sultanates to develop and flourish but perhaps making Central Asia vulnerable to the rise of the Mongols.

MONGOL DOMINATION

The Silk Roads, the paths of trade and merchants, became routes of terror when the Mongol Empire's armies dominated the roads, beginning in the 13th Century (Table 2.1). Genghis Khan, founder of the Mongol Empire, expanded Mongol territories through a series of strong military conquests. His empire did not end with his death in 1227, as his descendants and successive leaders continued their dominance. In 1258, Mongol armies, led by Hulagu, grandson of Genghis Khan, sacked Baghdad after a 13-day siege. To the victors went the spoils. The Mongols made off with gold, silver, gems, and pearls; no surprise there. But they also seized all of the silk they could gather, due to its value for establishing power among their armies. Accounts of the actions of the Mongols vary greatly, particularly the bloodshed caused by Mongol invaders. Numbers of citizens killed trying to flee range from fewer than 90,000 to nearly 1 million. Even at the low end of the range, the slaughter was swift and extensive. Buildings were burned, and some accounts told of the Grand Library of Baghdad being ransacked. Allegedly, so many books were dumped into the Tigris River that the water ran black with ink; the veracity of that tale is challenged by some historians. Regardless, it was many years before Baghdad returned to its former prominence.

As the Mongol conquests led to development of an empire, their invasions and fierce attacks caused justified fear, given the resulting devastation. Soon after the rise of the Mongols, Chinese monks who anticipated an invasion gathered more than 10,000 manuscripts and paintings made on silk and sealed them in a cave for protection. The extensive collection of protected items was discovered centuries later, in 1907, when the Caves of the Thousand Buddhas near Dunhuang were found and opened.

The ferocity of the Mongols produced peace. As various peoples were incorporated into the Mongol Empire, fighting among various groups diminished, replaced by stability. The period known as the Pax Mongolica, Latin for Mongol Peace, impacted the silk trade, the Silk Roads, and all of Asian civilization. During the Pax Mongolica, travel along the Silk Roads became safer because the network's roads were guarded against foreign invaders as well as thieves. That period of peace, lasting through the 14th Century, was instrumental in the growth of trade along the Silk Roads, expanding its scope and introducing new technologies. This period of peace and safe travel enabled the journey of the Venetian merchant Marco Polo. In 1271, Marco Polo accompanied his father and uncle on a journey that ended in eastern China after four years, where Marco Polo was put to service in the court of Kublai Khan. Marco served Kublai Khan for a number of years and was finally allowed to leave in 1292, arriving back in Venice in 1295. His homecoming was cut short when he was captured and imprisoned by rival Genoese. While imprisoned, he wrote his account of his travels, entitled "Description of the World." The tales of Kublai Khan's empire described a very foreign life, with its unique culture and innovations unknown in Europe.

After the death of Kublai Khan, the long spiral of decline of the Silk Roads began. Various other groups gained power. The Golden Horde, a mix of Mongol and Turkish people and influences, had its stronghold across the steppe. Ironically, the invasion of Tibet, which the monks feared, failed to materialize, but a different invasion happened; and this one happened because of the Silk Roads. Fleas that lived on great gerbils (*Rhombomys opimus*) or other rodents were transported adventitiously on the Silk Roads to trading centers and then to ports. What made that invasion so dire was that the fleas were carrying bacteria—*Yersinia pestis*—the causal agent of bubonic plague. The story is one of rodents, insects, bacteria, and human fatalities (Chapter 5). Rats are what most people remember about the story. There is much more to that story than the simple connection of species.

From the shadows of the Mongol Empire, and while the plague raged throughout Europe and Asia, a new leader emerged—Timur, also known as Tamerlane. Timur's reign from 1370 to 1405 included invasion of the Golden Horde in 1395, whose grip on the steppe territory was weakened by the plague. The former Mongol Empire and other captured territories were reconstructed as the Timurid Empire, with its capital at Samarkand. Timur used the spoils of his victories to buy the best available goods, including the finest silk made in China. The merchandise was said to have made its way to Samarkand on caravans of 800 camels. Timur was seemingly a person of paradoxes. Although he was a supporter of education and religion, and a patron of art and architecture, he and his troops are thought to have slaughtered more than 15 million of the people he vanquished. His military exploits took him all across Asia, and his death in 1405 occurred while he was on his way to start a campaign to extend his rule into China (Figure 2.10). Upon his death, with no clear successor, the empire dissolved.

Figure 2.10 The tomb of the Asian conqueror Timur, in Samarkand, Uzbekistan.
Source: Nicola Messana Photos/Shutterstock.com

NOT THE END—CHANGE

The decline of the Silk Roads was not due to any one event at any one time, but a number of events, spread over centuries. Travel along the roads was no longer safe after the dissolution of the Mongol and Timurid empires, and, as a result, the primary routes that formed the core of the Silk Roads became less important. That core extended to the west, to the east, and to the sea lanes that added to and linked the roads by the Maritime Routes (Chapter 3). The routes, both land and sea, shifted with time, avoiding enemies or including new trading partners, or they shifted as rivers literally dried up, forcing the opening of new routes. The emergence of new empires or conquerors changed the landscape, with some towns razed while other new centers were created. The primary routes along the Silk Roads are sprinkled with ancient monuments and historic buildings, many more of them lost to time, desert sands, and conquests. The Silk Roads were known for their routes of commerce. But the extensive networks of routes carried more than the commerce of the day, much more than valuable commodities.

The influence of the Silk Roads extended well beyond silk and well beyond the 1,500 years of their existence. In his book *The Silk Roads*, author Peter Frankopan identifies two dozen roads, many of which are still used today, and all of which contributed significantly to some aspect of human history.[3] The roads, traversing vast grasslands, skirting deserts, and following passes through the highest mountains in the world, introduced Eastern culture and religions to the Western world and introduced Western religion and culture to the East. Silk, gunpowder, and paper went west; horses and wheeled vehicles went east. The impacts on human cultures, economies, art, science, and history are enduring.

The Silk Roads didn't end. They exist today, albeit in a form that would have been unrecognizable to any of the primary players throughout their early history. Travel along the Silk Roads engendered a movement and mixing of people, leading to the exchange of ideas, of interconnected cultures, customs, and beliefs, of knowledge produced by different experiences but shared history. Travelers along the Roads were attracted to the cultural and intellectual fabric then being woven, with many of the major stops along the routes becoming centers of learning. The Silk Roads advanced science, mathematics, literature, art, languages, and religions, and became a singular force that shaped the diversity of societies and cultures across the continent and beyond. And all of this came about because of *Bombyx mori*, the little, blind, flightless moth, whose thin strand of salivary secretion—moth spit—led to a creation of a web of routes that knitted together East and West, passing thousands of kilometers east of the prime meridian and creating some of the richest chapters in human history.

NOTES

1. RNW traveled throughout Turkey in 2004 and one day was driven on a road that passed between the Tigris and Euphrates Rivers. We passed by farmers with large, round, flat, woven baskets, which were used to throw harvested wheat into the air.

Wind carried away the lightweight chaff, leaving the grain to fall to the ground, from where it was gathered. Agriculture was being practiced as it had been in its beginning. That area of Turkey was a series of paradoxes and contradictions: in very rural mountain valleys, there were low-rising houses made with a sod roof, but with a satellite dish attached to one corner; shepherds in vast montane grasslands were talking on cell phones while driving herds of sheep. The reach across millennia was best illustrated on a road where we witnessed a lone ox pulling a cart, the cart having solid—not spoked—wooden wheels. The oxcart was then passed at high speed by a Mercedes 500. There was at least 5,000 years of technological difference displayed right there in one moment.

2. P. Frankopan, *The Silk Roads: A New History of the World* (New York: Alfred A. Knopf), is nearly encyclopedic in its explanations and expositions of Eurasian history. Frankopan tells of the role that the Silk Roads played in the history of silk, but also much of the rest of human history over the past 2,000 years, as created or enabled by the Silk Roads. This beautifully written book is engrossing, and a reader using the book to find a page with a reference will not look until up after reading another 20 pages. It is *that* good.

   A second source for the complete history of the Silk Roads is Christopher I. Beckwith, *Empires of the Silk Road: A History of Central Eurasia from the Bronze Age to the Present* (Princeton, NJ: Princeton University Press, 2015). Also, another very thorough review (but not 512 pages) of the history of the Silk Roads is a chapter in a book on ancestral genetic makeup of populations of people: Rene J. Herrera and Ralph Garcia-Bertrand, "The Silk Roads," in *Ancestral DNA, Human Origins and Migrations* (New York: Academic Press, 2018).

3. Peter Frankopan lists 25 roads that succeeded or replicated the Silk Roads, such as "The Road of Faiths," "The Road of Death and Destruction," "The Wheat Road," and "The Road to Genocide." All detail significant aspects of history and all were at least partly in the region of the world where the original Silk Roads were located. The account of discussions preceding the German invasion of Russia is chilling in "The Wheat Road," and the profane explosion by Josef Stalin in "The Road to Genocide" tells more than we likely want to know about Stalin and the perils of telling a despot the truth.

CHAPTER 3

ᴄᴠᴐ

Silk Goes East and West

The name Silk Roads might conjure up a mental image of a string of Bactrian camels, lined up as a caravan, making their way across the high deserts of Central Asia. That works if your image is of the core routes of the Silk Roads from 100 BCE to about 1400 CE (Chapter 2). But that is just part of the story. The core of the route may have lost some of its importance, but the Silk Roads extended beyond that core in multiple dimensions—geographically, through different transportation modalities, and temporally, being vibrant, still, in the current millennium. In each dimension— spatial, modal, and temporal—the routes and the players changed. Extensions of the Silk Roads directly resulted from the network that evolved from silk production in Eastern China to the broad-scale trade of that silk that began more than 2,000 years ago. As silk went, so were the Silk Roads extended and expanded.

The Silk Roads had their parallels on the seas. Short-distance trade on land between nearby villages was akin to the short jaunts made by small boats from one coastal village to another. Longer land-based routes morphed into trade networks. By sea, the Maritime Routes were the oceanic equivalent of the Silk Roads (Figure 3.1) and formed a web of crisscrossing sea routes, connecting and complementing the land-based Silk Roads, linking all of Asia and linking Asia to the world. Connecting routes crossed mountains and passed along river valleys, taking wares to India, Arabia, and the Mediterranean. Silk Road branches passed through the Hindu Kush and followed the Indus River to the Arabian Sea. The Tea-Horse route, or the Southern Silk Road of southeast Asia, carried merchandise from Yunnan in southwest China to ports in the Bay of Bengal or to the southeast, to Vietnam, where village names along the Duong River included the word Dâu, that denoted the village's connection to mulberry trees, silkworm food.

Maritime trade emerged thousands of years ago, as traders passed by coastal cities, such as along the shores of the Red Sea or the Persian Gulf, or the extensive coastlines of India or China. Although most early trade vessels stayed close to land, as they plied their wares, trading occurred on short oceanic routes across the South China

The Silken Thread. Robert N. Wiedenmann and J. Ray Fisher, Oxford University Press. © Oxford University Press 2021.
DOI: 10.1093/oso/9780197555583.003.0003

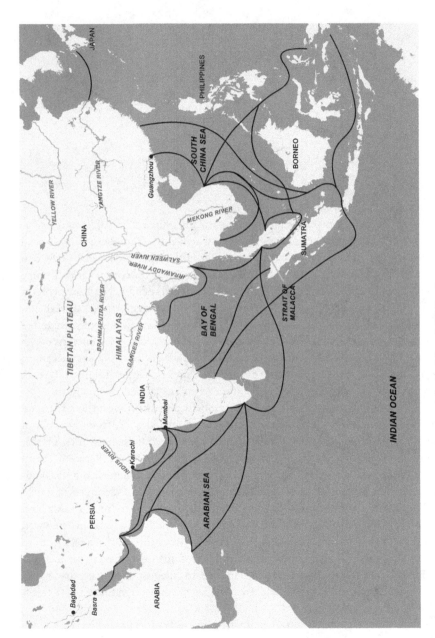

Figure 3.1 Maritime Routes that complemented the land-based Silk Roads.
Source: Sarah Silva, MapShop.com

Sea, between today's Vietnam and southeastern China, or across the Gulf of Oman connecting the Arabian Peninsula with the Indus Valley. The first known transoceanic trade occurred in voyages nearly 2,000 years ago between Oman and the Malay Peninsula, which held goods from China. Whether the fabled Sinbad existed or not, the voyages from the Arab world to the Far East opened up maritime trade.

Coastal trading cities popped up, such as Goa, Muscat, Zanzibar—exotic names in exotic lands. The cities became wealthy centers where not only were goods exchanged, but languages, ideas, and beliefs passed among people from diverse lands and cultures. The Chinese harbor city of Quanzhou was at the center of trade where goods were moved to and from the sea routes. Quanzhou and its prominence were cited by early travelers, including the Moroccan Ibn Battuta and the Italian Marco Polo.

EXPANDING THE ROADS

Expanding land-based travel required no real changes, in contrast to the changes needed to transform coastal trading into longer sea routes. Voyages of unknown duration required planning, to carry enough casks of water and provisions for the crew. Easily overlooked, voyages required dealing with nuisances, such as grain borer moths, whose larvae infested and ruined food supplies, or threats from shipworms— a group of wood-boring clams—whose boring into and chewing of wood fibers compromised ships' integrity. Development of hemp as a textile enabled making strong canvas sails (Chapter 1). Shipbuilding improvements produced vessels that could carry more weight and better endure oceanic conditions. The slight improvements made in newer maps, though they were still inaccurate, were a vast improvement over ancestral maps, which often reflected the imagination or wishes of a mapmaker and bore little resemblance to real land features.

Using the sun or stars to navigate was crude but, for eons, was all that sailors had. The compass changed that. Developed in China about 300 BCE, the compass was simply a piece of the mineral magnetite and was used initially for fortune telling or to ensure buildings faced a particular direction considered important to ensure a prosperous life for the building's inhabitants. The compass as an instrument, rather than a curiosity, was adapted for maritime use by the 11th Century, by magnetizing a needle and placing it on a piece of wood in a bowl of water. This crude "wet" compass enabled sailing vessels to safely pass farther from land. A further advance was a "dry" compass, in which a magnetized needle was mounted on a pin and sealed in a glass-lidded wooden box that could be secured to the ship, thus allowing for its use in rough waters. But the glass lid often darkened after extended exposure to the sun, making the needle difficult to see and, thus, the compass difficult to read.[1]

Other technology was added to the sailor's toolkit. The astrolabe proved valuable for calculating the positions of stars and, by the 11th Century, was used on ships to enable navigation even farther from shore and to predict landfall. Sextants arrived later, but they provided greater precision in determining a sailor's position on the featureless ocean.

The web that formed the Maritime Routes covered 15,000 kilometers, carrying valued goods from the "Spice Islands"—the original "Indies"—westward. The Spice Islands were among the archipelago of islands forming modern Indonesia, stretching 5,000 kilometers from Sumatra to West Papua, on the island of New Guinea. Trade along the Straits of Malacca and Sunda became prominent, ruled for centuries by the Indonesian Srivijaya Empire, which controlled sea trade throughout south and southeast Asia from the 7th to the 13th Centuries. Indonesian batik, made from silk or cotton fabric, added to the goods traded. Batik is made by hand-dyeing fabric, with areas covered by wax to resist the dye, thus creating combinations of color patterns. Spices were coveted in western Asia and the Mediterranean, then later throughout Europe. The contributions of spices to the trade earned these maritime destinations the title "Spice Islands."

The Maritime Routes, coupled with the specialized products that were carried on the Spice Routes, both complemented and competed with the trade that was occurring along the land-based Silk Roads. The Maritime Routes allowed goods to continue to move when traders along the land-based Silk Roads faced raids from marauding tribes—for example, the Mongols, who were at various times marauders and protectors. The sea routes went from east to west, but also in the reverse direction. In addition to carrying wares to trade, the Maritime Routes also took Islam to south Asia with the help of monsoon winds, establishing new centers of that religion, initiated by sailors who often waited for months for the reverse winds and, during their long wait, established second families. Today, Islam still maintains a stronghold in south Asia, with Indonesia being home to more Muslims than in any Middle Eastern country. And as the power of the land-based Silk Roads diminished, the Maritime Routes continued in use. All of that history, empire building, wealth accumulation, and power occurred because of silkworms and the product from their salivary glands. Silk moved east and west, but the knowledge of silk making was actively retained in China for millennia.

SILK GOES EAST

The history of silk—fabric of the Chinese emperors—is known better than the history of other textiles because of the historically tight control over the entire process. Every aspect of sericulture was under strict control—from growing and harvesting the mulberry leaves to growing the domesticated moths, collecting silk, and making fabrics. Sharing or distributing any information about silk production was, for centuries, an offense punishable by death. Those charged with keeping the secrets of the silkworms themselves, their production processes and harvesting, and the weaving silk threads faced conflicting demands. On the one hand, severe punishment was threatened for anyone who passed along the secrets. On the other hand, there was a potentially great economic benefit for anyone to pass along information about growing silkworms and moving the process elsewhere.

Silk's secrets remained in China until a large number of Chinese emigrated to Korea, taking their knowledge of silk making with them. The date of that emigration is not clear, as various sources put the date for silk production arriving in Korea somewhere from as early as 1200 BCE to as late as 200 BCE. What is known for certain is that the secrets of silk and silk making were passed along from Korea to Japan about 300 CE (hereafter dates are assumed to be Common Era unless stated otherwise).

At the far eastern end of the Silk Roads, a short sea route crossed the Sea of Japan from Korea to Japan. The city of Nara, which became the capital of Japan in 710, was linked to the Roads through the coastal city of Osaka. In Nara, the cultures of Japan, China, and Korea merged and showed the evidence of influences drawn from the entire length of the Silk Road. Buddhism had already reached Japan by the end of the 6th Century, having made its way from south Asia. Nara's influence lasted until 784 and, at its zenith, the city was a religious center, with the large number of religious buildings and temples built during that time showing the prominence of Buddhism. The Great Eastern Temple contains the Great Buddha Hall, one of the world's largest wooden structures, holding within it a 15-meter tall bronze statue of a seated Buddha.

Nara also produced a number of goods that were traded along the Silk Road. Jewelry, musical instruments, sculptures, and silk paintings influenced—and were influenced by—artwork all the way to Western Asia. For example, Buddhist art from the Nara period has been found in areas as isolated from each other as Indonesia and Afghanistan. Specialized goods from Japan, such as metalwork and carved wooden pieces, were transported across the Sea of Japan and added to the caravans that went west. In the opposite direction, artwork from the west was carried over the Silk Roads to Japan, where it was desired. Designs that included dragons were similar to ones created by the Scythians in northern Steppe lands. Curiously, a bronze crown located in Japan is nearly identical to one found in Afghanistan, at the far western end of the Silk Roads.

Silk, too, was part of the mix, once its production was established in Japan. A specialized silk-weaving technique, called Yuki-Tsumugi, used floss from deformed silkworm cocoons and a special dyeing process to produce a light, desirable, silk fabric. The silk that was produced in Japan not only included products from the imported *Bombyx mori*, but other products from the native Japanese silk moths (*Antheraea yamamai*). Japanese silk moths belong to a different family than domestic silk moths (Bombycidae), called giant silk moths (Saturniidae). Giant silk moths, like most moths, also produce silk. Silk from Japanese silk moths has been prized since the 11th Century, and still today the strong, white silk is uncommon and expensive. Curiously, the silk produced by Japanese silk moths is mostly resistant to dyeing techniques. Unlike domesticated silk moths that feed nearly exclusively on leaves of white mulberry (Moraceae), Japanese silk moths feed on leaves of a number of trees—oaks, chestnut and beech (Fagaceae), hawthorn and rose (Rosaceae), and a few others.[2]

SILK GOES WEST

The Chinese had kept the secret of silk making nearly intact for millennia. Although silk fabric had made its way west by early in the first millennium CE, the underlying secrets of silk and its production did not travel with it to the Roman Empire. Early ideas as to silk's origin included silky covering on leaves, petals of flowers from trees, even certain types of soil. The Romans had no idea what silk was—let alone how it was made. Despite not knowing what silk was, the Romans were fascinated by this new fabric. In fact, fascinated would be an understatement, obsessed might be the better descriptor. It was a case of the public running ahead of their leaders. The desire for silk was nearly impossible to meet, and silk was worth its weight in gold.

Asian goods that arrived in the Roman Empire were considered decadent and counter to the empire's self-proclaimed virtues. Virtues? Recall the stories told of the Coliseum and cheering Roman crowds witnessing Christians pitted against lions or gladiators fighting to the death. Decadence? This was an empire that was led by Caligula from 37–41 BCE. Although modern interpretations have cast doubt on the veracity of tales regarding the worst of Rome's ills, regardless, according to the Roman leaders, goods from the East were decadent and nothing epitomized that status more than silk did. That decadence had real economic consequences. At one point in the 1st Century, spending on silk was equal to more than 10% of Rome's annual budget, and that money was sent out of the empire, thereby harming the Roman economy. According to Pliny the Elder, spending on overpriced silk simply to "enable the Roman lady to shimmer in public" was not good for the empire. But laws and decrees failed to defeat public opinion and failed to quell the desire for silk.

Roman leaders decreed import limits by regulating prices, with only limited success. They tried to enact laws to prevent men from wearing silk, trying to convince them that wearing expensive fabrics imported from Asia was disgraceful. Think that worked? Ha! Then in the year 14, Emperor Augustus issued an edict that women could not wear silk gowns. In what was apparently the first-known political action by women, they gathered together to lobby for the edict to be overturned. And it worked! Silkworm and women's rights.

With time, silk was worn not just by Roman nobility, but simple silk garments were being worn by common people. However, the finest silk was still out of reach of most commoners. The price for some types of fine silk clothing was exorbitant, with the cost some very fine silk being equal to a Roman soldier's *annual* salary. As more women began wearing silk robes, their attire caught the attention of many, with the philosopher Seneca the Younger capturing in the 1st Century the essence of concern, stating that the silk clothing was indecent. It seemed to him—and many others— that the gowns worn by women left nothing to the imagination.

One reason for the high cost of silk in Rome was the distance over which the precious fabric had to be carried to reach the empire. The Parthians (Chapter 2, Table 1), who exacted revenue from the traders by serving as middlemen and charging the traders to pass through their lands, increased the costs paid by the Romans. The supply-demand axis was so strong, and the monetary benefit to the Parthians so great, that the mortal enemies avoided aggression. An insect as a peace broker?

Thank you, silkworm! One other result of the Parthians' status as intermediaries was that there was little to no direct contact between the Han Dynasty and Roman Empire. The Han and Romans were connected by the silken thread, but so indirectly through the Parthian middlemen that neither knew much about the other.

SILK-MAKING MOVES

Because of high demand for silk and the amount of money to be made from producing and trading it, escape of the secret from China was inevitable. The first escape of the knowledge needed for silk making had already occurred when Chinese emigrés settled on the Korean Peninsula about 200 BCE. Knowledge of silk production found its way to India about 400 years later, then on to Mesopotamia (modern Iraq), and then to Persia (modern Iran), where silk making was established and flourished. Eventually, silk found its way to the Byzantine Empire.

Mesopotamia was a center for trading, and Baghdad was the primary central market. Although each of the major cities was famed for its particular type of woven silk, Baghdad was the major silk center in Central and Western Asia (Figure 3.2). Other materials, such as cotton and wool, were woven with the silk to create different styles of fabrics, but Baghdad was known for the gold threads that were woven into the silk or were used for embroidery, and silk produced locally in Baghdad was even exported eastward along the Silk Roads. Centuries later, the methods of Iraqi silk weaving were copied and used in European silk manufacturing.

Figure 3.2 Persian silks displayed in a market along the Silk Road in Uzbekistan, Central Asia.
Source: Iryna Hromotska/Shutterstock.com

One interesting aspect of Mesopotamia and the Silk Roads was trade of a different product associated with China—not silk, but porcelain. Fine Chinese porcelain arrived by the Maritime Route in the Iraqi port city of Basra by no later than the 9th Century, inspiring local pottery artists to try to copy the style. Chinese porcelain was made with kaolin clay and, when a clay piece was fired at very high temperatures, the process yielded a semi-translucent white ceramic piece. However, the yellowish clay available to potters in Basra did not contain kaolin. Faced with the limitation of fine clay, the potters used the yellow clay, but they developed an opaque white glaze to use as a base and onto which they created brilliant blue and white patterns. In time, as the new porcelain from Basra was traded, it found its way back to the East, where Chinese potters tried to emulate the porcelain from Basra. In their efforts to imitate the Middle Eastern porcelain, the Chinese created even more beautiful blue and white porcelain, and the back and forth continued. Both the finest of the ceramic wares and the processes to create them spread throughout Western Asia. Whether the Basran pottery inspired the Chinese porcelain of the Yuan (1279–1368 CE) and Ming (1368–1644 CE) dynasties is still not decided.

The Persian Empire created a strong and enduring center of silk production, beginning with the rise to power of the Sasanians in 224 CE. At first, the Sasanian weavers disassembled woven textiles and reused the silk. Later, they imported raw silk from China along the Silk Roads, allowing them to create new and brilliantly colored fabrics. Once in Persia, the silk was dyed and woven into elaborate patterns of brocades and twills distinctive to the Sasanians. The primary design showed a repetition of a pattern consisting of a round central space containing a pair of animals, often lions or stags, faced off as if in confrontation. Surrounding the paired animals was a ring of complex design, termed a roundel, and the repeated roundels alternated with floral patterns. The well-known and desirable silk textiles were then exported eastward back to China or west into Europe. In many places the textiles were exported to, wearing elaborate Persian silks, especially those with gold interwoven threads, was limited by law to nobility or religious figures. In time, these Persian patterns with their complex designs and symbols were copied by weavers across Eurasia, influencing art and other uses of the designs. Persian fabric artists had in their repertoire beautiful patterns and complex weaving techniques, but they still needed to import the raw silk—production of silk was limited to China and India.

SILK INTRIGUE

Neither the Han Dynasty, the Parthian, and later the Sasanian Empires, nor the Western Roman Empire survived past the middle of the first millennium CE. The Han Dynasty, considered the golden age of Chinese history, was succeeded by a series of short-term rulers, and not until the Tang Dynasty in 618 CE was there a return to great power in China. The Parthians were replaced by the Sasanians, who ruled for more than four centuries as a world power—at least the world that revolved around the Near East and Middle East. The Roman Empire fell to Germanic forces in 476 CE, though the empire continued as the Eastern Roman Empire, also known as the

Byzantine Empire, with its capital in Constantinople—the important site where Europe met Asia and the western terminus of the Silk Road. The Byzantine Empire persisted for nearly a thousand years.

With the rise of the Byzantine Empire came the rise of silk making within the empire, though raw silk still had to be imported for the weavers. That changed in the decade of 560 CE. The often-told tale of silk's "escape" to Constantinople—perhaps more apocryphal than truthful—centers on two monks and their travels and travails on the fabled Silk Road. Parts of the tale of smuggling silkworms to Constantinople seem to pass the test of veracity, though a few questions remain.

In approximately 550 CE, two Nestorian monks were preaching in India from where they traveled to China. At that time, some westerners believed that silk originated in India. When the monks visited China, they realized they had found the true ancestral home of silk. While in China, they observed silkworms, how they were fed mulberry leaves and how silk was harvested.

According to legend, the monks returned to Constantinople and asked for an audience with Justinian I, the Byzantine Emperor. They explained that silk was indeed from China and was produced by an insect. They described the insect and its life cycle, its reliance on mulberry leaves, and the possibility of transporting silk moths back to the empire. They told Emperor Justinian that by producing silk themselves, Byzantine weavers could create their own silk industry because they would no longer need to buy silk from their enemy, the Persians. The monks stated that they could provide the materials to produce silk. Convinced, Justinian ordered them to return to China for the purpose of smuggling silk moths. Recall, the Chinese had a ban on exporting silkworms of any life stage or the knowledge of how to produce silk, under penalty of death. The monks must have escaped scrutiny, as they were able to carry back with them knowledge and skills for growing silkworms, their product—silk— and the silkworm eggs to start silkworm and silk production.

Here the story gets a little fuzzy. By one account, the monks were able to smuggle eggs in their hollowed-out bamboo walking sticks. Eggs are less affected by harsh conditions, such as aridity or freezing temperatures, and so this story is plausible. However, the return trip would have taken nearly a year, and it is a stretch to think that the eggs were insulated sufficiently from temperatures ranging from bitterly cold at night to baking under a blazing sun. So, maybe.

Another explanation recognizes the unlikelihood of eggs surviving that long, and so the second tale suggests that the monks made the return trip in stages. To do so, they would have needed to stop periodically in oases to allow the eggs to hatch and the larvae to feed long enough to complete their life cycle, go through pupation, and emerge as adult moths. Adult moths then would produce another generation of eggs, which were repacked to survive the next leg of the journey, and the monks would set off again. Transportation would have required moths to be in the egg stage, because eggs are less prone to desiccation and are more easily insulated from temperature changes. Again, maybe, but perhaps this story is more plausible.

However, this story has a major flaw as well. It begins with the Nestorian monks traveling from India to China, where they realized the true source of silk. Continuing, they brought this knowledge to Justinian, who apparently was not aware that silk came from China. However, as far back as the 1st Century CE, the Greeks and Romans referred to the area of East Asia corresponding to northern China as "Serica" and its people as the "Seres," meaning "where silk comes from." The word silk in Latin is *sericum*, and the culture of silk moths is called sericulture. It is a stretch to assume that 700 years later the emperor of the Byzantine Empire was unaware of the source of silk.

Yet another tale of the smuggling of silkworm moths from China is told as a brief love story. In this legend, the Prince of Khotan—an oasis town on the route north of the Taklamakan Desert—fell in love with a Chinese princess. As part of her dowry, she allegedly smuggled eggs from the silk moth in a large hairpiece. If so, that subterfuge resulted in moving the eggs only as far as Western China, leaving the remaining movement to occur by unknown means. Did she really allow moth eggs to be put into her hair? Why did the eggs not hatch?

Whatever the explanation, entomological questions remain: How did the monks know enough about insect biology to know that eggs could be insulated and kept from dehydrating, freezing, or hatching? How did they learn this, if sharing information in China could result in the death of a silkworm farmer or the deaths of the monks themselves? How could they even ask about insulating eggs without tipping their hand? If the multiple-stage explanation is valid, what did the monks know about the silkworm moth's life cycle or how long eggs could be kept before they died (or hatched)?

There are questions about the plants, too. White mulberry did not occur naturally in Byzantine Asia in 550 CE, so the traveling monks would have had to know that oases along the route had growing mulberry trees. Or, perhaps, they would have needed to carry live plants or mulberry leaves with them to be fed upon by silkworms. They would have needed to keep the plants alive or the leaves fresh, regulate how much the larvae fed on, and have viable plants awaiting them upon reaching Constantinople. Further, they needed to *smuggle* the plants, for there would be no reason to transport mulberry plants just to say that they would look nice back in Constantinople. The monks would have needed to know enough about silkworm-plant interactions to return with both living plant and insect specimens. Again, maybe this tale is true—but only maybe.

SILK IN THE BYZANTINE EMPIRE

All the stories about and the explanations of the smuggling of silk out of China are a bit fuzzy. Nonetheless, silkworm adults or eggs or larvae ended up in Byzantine Asia and production of silkworms and silk soon followed. The life cycle of the silk moths was recreated and feeding on mulberry leaves by silkworm larvae continued as it had for centuries in China. Intrigue, deception, and disbelief aside, silkworms and silk production actually got to Constantinople—and they did so 1,500 years ago!

Prior to the arrival of silkworm eggs in Constantinople, raw silk had to be imported from China. Regardless of how silkworms got to Constantinople, the Byzantine Empire was in the silk business by the middle of the 6th Century. Once local silk was produced, weavers could produce silk fabric, avoiding the high costs of importation. At that point, local production increased quickly and Constantinople, at the intersection of Europe and Asia, became a significant silk producer. Silk production and weaving were delegated to women, just as had occurred in China, in "factories" called gynaecea. Silk, as currency and as a form of portable wealth, brought great riches to the empire and enabled trade for desired goods and revenue to support it. The outcome of all the efforts was a new silk industry in the Near East, one that endured for centuries.

Byzantine silks were prized for their rich colors, with a variety of dyes used as well as gold-wrapped threads that gave the silk textiles and clothing a luster unmatched anywhere else. Modern looms and techniques were used to improve upon earlier weaves, such as twill from Persia, and to create totally new weaves, notably damask. One weave was a compound, heavy twill, called samite, which included gold or silver thread woven in to give a polychromic sheen. The dye materials used to create fabrics colored purple were tightly regulated and, so, silks dyed purple were reserved for royalty or to be presented as gifts to leaders and diplomats from other lands. The silk robes created by Byzantine weavers set the style for religious and military figures beyond the empire. Over the centuries of Byzantine silk production, both the patterns and techniques in Byzantine silk weaving evolved, reflecting ongoing exchanges with textile centers in Central Asia after its conquest by Islam.

As their predecessors in China, India, and Persia had done before them, Byzantine silk producers monopolized silk production, and they worked to keep the secret to themselves. Silk and finished silk products were transported along the Silk Roads despite the continual wars between competing empires. Silk was worth its weight in gold—in places replacing gold as currency for payment or as gifts—so the need for it demanded finding a way to avoid the wars. In some cases, safe travel was permitted if the traders paid for their safety, thus increasing the cost of the goods considerably, even doubling the price. Or the trade goods were confiscated, and the traders were killed—after all, again, the confiscated silk was worth its weight in gold. Local strife sometimes resulted in the use of less-direct routes, such as the steppe routes via the Black Sea, development of new routes, or finding totally new areas for trade, such as westward into Europe: Byzantine textiles and silk robes made their way to cathedrals throughout Europe.

Byzantine weavers transformed silk and silk transformed the Byzantine Empire. Silk, and its use in rituals, transformed cultures from the Islamic world to Western Europe. The product of the silkworm larva was the source of intrigue, treaties made and treaties broken; it was exchanged for military aid and was the economic engine that paid for the empire's exploits. However, even with its great success and impacts, the Byzantine silk industry did not last. In 1147 during the Second Crusade, two centers of Byzantine silk production, Thebes and Corinth, were captured. In other wars, the captors seized weapons or gold coins. In this instance, what was taken were the skilled weavers and their looms, transported to Italy for to create a new silk industry. During the Fourth Crusade in 1204, Constantinople was besieged and then ransacked, marking the end of the monopoly, as silk making moved to Italy.

The conquest of Constantinople and fall of the Byzantine Empire at the hands of the Ottomans in 1453 led to a stop in the flow of goods along the roads. At the time, the Byzantine Empire was already greatly reduced in scope and power, as silk making had been moved to Italy and France. The population of Constantinople—as throughout Europe—had been decimated by the plague, which had been carried west on the Silk Roads along with goods for trading.

By the end of the Middle Ages, as silk making was on the rise in the European continent, the connections to the Silk Road were mostly lost. With all the preceding pressures and losses, it would not take much more for the Silk Roads to come to an end. Probably the proverbial straw that broke the camel's back (Bactrian, of course) of the Silk Roads was the discovery of transoceanic sea routes that expanded the network of the Maritime Routes and proved to be far more efficient for trade. The Age of Discovery, with faster sea routes plied by sailors from different European empires, brought the world together in a manner similar to the original Silk Roads.

SILK GOES TO EUROPE

Silk and silk making didn't end with the demise of the Byzantine Empire. To the victors in the Crusades went the spoils, and the spoils included silk making and the infrastructure to support it. Once silk was in Italy, and with imports from China all but gone, silk making flourished to satisfy the demands for luxury silk. In the late 15th Century, there were more than 7,000 weavers and other crafters in Florence alone. Silk making developed differently in the city-states of Florence, Genoa, and Venice, each becoming known for a particular silk specialty and exporting silks throughout Europe, most of which were expensive and out of the reach of commoners. Improvements in looms allowed the creation of more intricate fabrics and designs, but the improvements added to the cost of the garments produced. However, the technological advances in the textile industry produced advances in other technologies, which in turn promoted further advances in the textile industry, such as new looms or water-powered mills.

The French assumed the dominant role in silk production in the 16th Century. In 1466, King Louis XI of France had wanted to develop a silk industry to produce luxury textiles, but his efforts produced minimal results. Nearly a century later, Francis I granted a monopoly to the city of Lyon to produce silk, and the Lyonnais became the continent's leading silk producers, creating a distinctive style. The economic impact of Lyon's silk industry was measurable—reportedly, there were more than 14,000 looms and 28,000 registered silk workers in the city.

With the Edict of Nantes in the late 17th Century, countless Huguenots, seeking to avoid religious persecution, emigrated from France to England. Many of those were weavers and sericulture experts who took with them their weaving and silk-producing expertise. However, the British climate was not conducive to growing mulberry, so the silk trade never became as dominant there as had the French industry.

But the silk industry on the European continent was not without its own problems. Beginning in Italy at the turn of the 19th Century, silkworms in culture for production were becoming sick and dying. The disease was contagious and soon had spread among silkworms far and wide, threatening the silk industry. In 1807 in Italy, bacteriologist Agostino Bassi began studying silkworms and after 25 years was able to conclude that the disease was caused by a fungus, transmitted by both infected food and contact with other infected individuals. He also developed methods to prevent and eliminate the disease. Bassi's work on silkworm disease represented the formulation of the first germ theory of disease. He is seldom acknowledged for his role in saving the silk industry, although he is remembered in the name of the fungus, *Beauvaria bassiana*. Instead, the person that gets the accolades as the savior of the silk industry was the French scientist Louis Pasteur.

Pasteur was originally a chemist, and his early contributions to science were studies of the crystals of different compounds. What didn't Pasteur do? He figured out fermentation and the process that bears his name, pasteurization, to avoid spoilage. He disproved the idea of spontaneous generation and later turned his knowledge and energy toward the study of diseases of humans and other animals. But in the mid-1800s, the French silk industry was under siege from yet another disease—not the fungus that Bassi had addressed. Pasteur took up the study of silkworm disease and found that it was actually two diseases—one called pebrine, caused by a microsporidian (*Nosema bombycis*), the other flacherie, a bacterial disease (though now believed to be caused by a virus). Pasteur's contributions, in addition to demanding standards of hygiene to reduce accidental infections, were to establish uninfected lines of silkworm, to eliminate hereditary diseases. Females oviposited and the eggs were examined: if infected, they were destroyed; if uninfected, the eggs were placed on mulberry leaves and silkworms were grown to adulthood. Pasteur's method resulted in only uninfected silkworms being kept for breeding, eliminating infections from both pebrine and flacherie. Pasteur's contributions now seem so obvious—they were not at the time. Ironic, too, is that Pasteur's germ theory was not initially accepted by the medical establishment, because it was developed by a chemist.

THE JACQUARD LOOM AND ITS IMPACTS

With the contributions of Bassi and Pasteur, the silk industry was on stable footing. The innovations generated by the Industrial Revolution benefitted the textile industry, with advances in the cotton industry made through progress in spinning. The silk industry benefited more from advances in weaving. Punch-card looms made their appearance in 1775, and improvements made by Joseph-Marie Jacquard produced the Jacquard loom at the turn of the 19th Century. Jacquard's machine used a series of connected punch cards to be processed in a defined sequence, producing complex and detailed patterns that could be made repeatedly, enabling mass production (Figure 3.3).

Several different outcomes ensued from introduction of the Jacquard loom. First, prior to introduction of the punch cards, complex silk patterns required a master weaver paired with an assistant, called a draw boy, who sat on top of the loom and raised and lowered warp threads manually. The new loom allowed for the production

Figure 3.3 Jacquard loom showing chain of punch cards.
Source: Gorosan/Shutterstock.com

of quality woven textiles by an unskilled worker in less time, thus reducing the number of workers needed.

Second, the reduction in numbers of skilled workers resulted in the workers, known as canuts, being employed in poor working conditions and paid poor wages for their products, leading to worker revolts. The initial uprising in 1831, when the workers took control of the silk district of Lyon, was the first known worker revolt. The second uprising in 1834 lasted six days and was quelled by 12,000 armed French soldiers, resulting in the deaths of 300 rioters.

Third, the need for fewer workers and faster production led to lower costs for mass-produced textiles, which were then affordable to a broader market of consumers.

And the fourth outcome affected the world greatly and still affects you today. The punch card that enabled quick production of silk patterns was a technology that found other uses, one of those extending the silken thread across the Atlantic to the United States. Herman Hollerith (Figure 3.4) graduated from Columbia University School of Mines in 1879 and was employed the following year by the US Census Bureau. The Census Bureau needed a way to tabulate census data better and more quickly than the hand entries and counting that had been used previously. Hollerith, employed as a statistician, followed up on an idea of a coworker to create a device similar to the Jacquard loom to automate the count.

Hollerith designed a machine to punch holes in paper cards, with the location of the holes on the cards indicating individual characteristics, and another machine to tabulate the information on the cards. His machine was tested on the collected 1880 data, yielding results that closely matched the hand-counted data, close enough that he was granted a contract for the 1890 Census. He was granted a PhD for his

463-465 PENNA. AVENUE.
WASHINGTON, D. C.

Figure 3.4 Herman Hollerith (1888), inventor of punch card and reader system.
Source: Charles M. Bell/U.S. Library of Congress/Public domain

invention and a medal at the 1893 Chicago World's Fair. Three years later, Hollerith founded the Tabulating Machine Company, based in Washington, DC. He was again granted a contract, this time for the 1900 Census. Hollerith believed he alone could provide the technology, so he greatly increased his prices, resulting in the Census Bureau developing a counting machine for their own use, which the government then patented. While in a legal battle over patent infringement, Hollerith sold the company, which then became part of a conglomerate called the Computer Tabulating-Recording Company. Hollerith maintained stock in the new company and served as chief consulting engineer.

The 1914 hire of a new company director proved fortuitous. Thomas J. Watson led the company as it improved both the tabulating machine and the sales of it and its applications, setting the company on stable footing again. Hollerith retired in 1921, and in 1924, Watson changed the name of the again-successful company to the International Business Machines Corporation, later known as IBM.[3] Yes, that IBM. Your computer, laptop, tablet, and cell phone are descendants of the card-reading tabulating machine that was patterned after an automated loom for weaving silk, that silk being the product of the salivary glands of the domesticated moth, *Bombyx mori*. The next time you log on or answer a text, think of silk, the silken threads made from moth spit.

NOTES

1. In 1979, RNW was with lifelong friend David Enstrom on a small boat used for cod fishing south of St. John's, Newfoundland. We had hired a local fisherman, John Reddick, to take us to offshore sea stacks to look at seabirds, such as kittiwakes, murres, razor-billed auks, and Atlantic puffins. The wooden boat was maybe 4–5 meters long, with high gunwales. As we made our way among the sea stacks that bright, sunny morning, John remarked that the sun had darkened his compass, making it hard to read and that he needed to buy a new one. We didn't understand the need—the sun was out and we could see land nearby. As he took us around the stacks, we barraged him with a string of questions about cod fishing. John ran a small fishing enterprise, which consisted of himself as the boat's owner and captain along with his brother and son. He invited us to accompany them that afternoon as they retrieved their nets. We set out in the sunny afternoon, with a small trailing boat used to pull in the nets containing cod, which were dumped directly into the boat in which we stood, knee-deep in cod. Once the nets were emptied, collected, and placed in the trailing boat, we set off for shore.

   David and I had not noticed how foggy it had become, foggy enough that we could barely see the trailing boat and could not tell which way shore was. Now the darkened compass came into play as John squinted to try to make out the needle. We tried looking and could discern even less of the needle. John's brother sat in the bow, allegedly to watch for rocks, though he frequently took a long swig of "Newfoundland Screech" rum, and he would shout and wave his arms wildly as rocks emerged through the fog. David and I were certain we would crash into unseen rocks and never set foot on land again. Almost certainly, John knew exactly where he was based on the colored buoys denoting where known individuals had lobster pots, or other fishing areas, signposts that we could not decipher. He probably wanted to put a little fear into these "city boys." But we learned about the value of being able to see a compass needle under sun-darkened glass.

2. In a perhaps not surprising parallel, the Japanese silk moth has been introduced to Europe for production of the "tussar silk" for which it is known, much as gypsy moth was introduced to North America to create a silk industry. Japanese silk moth, like gypsy moth, feeds largely on oaks but also on many more plants, Japanese silk moth has established in the wild and is spreading through Europe—just as gypsy moth spread across North America.

3. *IBM and the Holocaust: The Strategic Alliance between Nazi Germany and America's Most Powerful Corporation* (New York: Crown Publishers, 2001) by Edwin Black argues that Watson directed the company's collusion with Nazi Germany in their commission of genocide. Black provides detailed evidence (expanded in a 2012 edition) of the use of punch cards and card-reading machines for the worst of evils—to organize and streamline the deportation and decimation of Jews in Poland and Eastern Europe.

SECTION 2

Oriental Rat Flea and the Plague

—476 CE   Western Roman Empire ends
—500 CE   Middle Ages begin
—527 CE   Rule of Byzantine Emperor Justinian begins
—536 CE   Volcanic eruptions cause crop failures and famine
—540 CE   First Pandemic begins in Egypt, begins in Constantinople in 541
—1331 CE   Second Pandemic begins in Mongolia, moves along Silk Roads
—1341 CE   Siege of Caffa, plague moves to Constantinople
—1347 CE   "Black Death" begins in Europe
—1382 CE   *Quaranta giorni*—40-day quarantine initiated in Venice
—1665 CE   Plague of London
—1855 CE   Third Pandemic begins in Yunnan Province of China
—1894 CE   Plague arrives in Hong Kong, Yersin and Kitasato discover cause
—1896 CE   Plague reaches India by Maritime Route
—1898 CE   Simond finds infected fleas on rats in India
—1903 CE   Charles Rothschild describes *Xenopsylla cheopis* (Oriental rat flea)
—1914 CE   Bacot and Martin discover plug in flea foregut, describe disease transmission
—1925 CE   Geneva Protocol enacted—United States and Japan do not sign the accord
—1930 CE   Soviet Union Anti-Plague Institute expands into Central Asia
—1931 CE   Japan invades Manchuria
—1938 CE   Japanese Unit 731 biowarfare program begins
—1939 CE   Insecticidal properties of DDT discovered

CHAPTER 4

⌇⌇

In Reverse Order—The Third Pandemic First

In Chapters 1–3, we discussed silkworm moths (*Bombyx mori*), the product they create from their salivary glands, and the silk textile we create from that. We hope you recognized that the greatest benefit of those moths was not silk itself, but the complex web weaving together human cultures and altering the course of history. A simple thread connects silkworms, mulberry trees, and humans. Snip the link between any two and the entire thread is broken. No mulberry? No silk. No silkworm? If there is no silkworm, then humans would have no reason to value mulberry trees. This simplified example includes only three species, but insects are connected to other species by threads that we tend to recognize only when they affect us.

The four chapters in this section discuss another such web of connections, this one among several mammals, including gerbils, rats, and humans. Those are known as megafauna and occupy the top tier of organisms, because of their size. The next tier is much smaller—fleas and lice, for example, are only a few millimeters long. Then there are single-celled organisms—amoebae and bacteria. We don't see them, usually, but that doesn't mean they're not important. The one bacterium discussed in this section—just one—matters quite a bit because it causes a deadly disease.

But that disease only occurs when all the other organisms in the web are connected—by the tiniest of threads. Remove one species and the process stops short. And when a disease occurs, it may escape notice until the scale of the disease enlarges. The continuum of an infectious disease can increase in scale as it progresses from an incidence in one or a few individuals, to an outbreak (a sudden rise in the incidence of a disease), to an epidemic (when the outbreak spreads quickly, affecting many individuals at one location), to a pandemic (when an epidemic has spread over a broad area and infects a large number of individuals).

This section covers the continuum from incidence to pandemic of a disease known as the plague, which is transmitted by insects but is influenced by a web of species

The Silken Thread. Robert N. Wiedenmann and J. Ray Fisher, Oxford University Press. © Oxford University Press 2021.
DOI: 10.1093/oso/9780197555583.003.0004

connected by a silken thread. There have been three plague pandemics, each slightly different. The silken threads that run through the three pandemics weave patterns that are not readily discernable. The complex stories include numerous connected organisms—for each pandemic, remove one connection and there would have been no pandemic. To best explain the histories, we begin with the most recent pandemic. That story requires both geographical and biological contexts, which is where we begin.

THE BEGINNING

The Tibetan Plateau, also known as "The Roof of the World," is a high-elevation region of Asia encompassing several provinces in southwestern China, northern India, Tibet, and Bhutan (Figure 4.1). It is bordered by great mountains: the Karakorams, Pamirs, and Hindu Kush to the west, Himalayas to the south, the Kunlun to the north, and the Qilian to the northeast. North of the Kunlun lies the Tarim Basin and the Taklamakan Desert, themselves bounded further to the north by the Tien Shan. The Kunlun and Qilian Ranges separate the plateau, on the south, from the Great Silk Road running just to the north. The plateau is a geological paradox, as its elevation and location make for an arid region, receiving less than 300 millimeters of rain per year, but it also is home to tens of thousands of glaciers with locked up precipitation whose annual melt produces water used by one-third of the world's people.

Located at the eastern edge of the plateau, China's Yunnan Province figured prominently in early recorded history. Running through the province was a southern spur of the Silk Road, connecting Yunnan with India, known as the Tea-Horse Road (Figure 4.1). Tea from Yunnan was traded for Tibetan horses, which were valued by early emperors to the north. With time, Yunnan also became linked to the world through its rivers, which joined the Maritime Routes, themselves complementing the network of Silk Roads.

The major waters that drain the Tibetan Plateau flow through Yunnan (Figure 4.1). The far northwest of the province is marked by the Hengduan Mountains, major mountains themselves, with many peaks exceeding 6,000 meters. This north-south range is divided by deep river valleys that drain the plateau. Three of these valleys carry the major waters of eastern and southeast Asia—the Yangtze, Salween, and Mekong Rivers. The Yangtze, the third-longest river in the world and perhaps the most important river in China, turns east as it leaves the Hengduan to head toward the East China Sea, which it meets at Shanghai. The Salween is perhaps best known for the Burma Road, the road to Mandalay, which runs alongside it and was strategically important in World War II. From Yunnan, the Salween goes south, passing through Myanmar (formerly Burma) and along the border with Thailand before emptying into the Andaman Sea in the Indian Ocean. And the Mekong River, known for its role as a route for supplies and soldiers in the Vietnam War, crosses borders

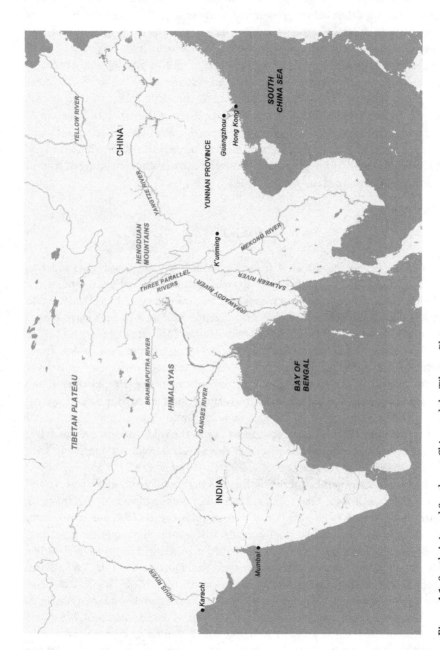

Figure 4.1 South Asia and Southwest China and the Tibetan Plateau.

Source: Sarah Silva, MapShop.com

with Thailand, Laos, Cambodia, and finally, Vietnam, where the river forms the great Mekong Delta and meets the South China Sea south of Ho Chi Minh City.

As the three great rivers leave the Tibetan Plateau, they even pass within a few kilometers of each other before proceeding on their way. The Three Parallel Rivers Protected Area in northwestern Yunnan Province, where the rivers converge, is considered perhaps the most biologically rich and diverse temperate area in the world, the reason for its protected status. More than half of China's plant species are found in Yunnan, with an estimated 18,000 plant species found in the province. The Three Parallel Rivers Protected Area alone is home to more than 6,000 plant species, at least 400 species of birds, and 173 species of mammals; many of those plants and animals are endemic to the rich region of Yunnan. Among those mammals are a number of rodents: marmots, mice, and especially rats. All of these features contributed to the generation and spread of the plague and, therefore, to our understanding of it. And we have not yet mentioned insects.

RODENTS

Of the 6,000 species of mammals worldwide, about 2,400 are classified as rodents. Rodents are highly successful, living in diverse habitats on every continent except Antarctica. The name "rodent" may mean "rats and mice" to you, but other familiar rodents include squirrels, guinea pigs, and porcupines. Some rodents have names that conjure up images of Dr. Seuss characters: the thick-tailed three-toed jerboa or the false zokor. Some rodents have adapted to living near humans or our dwellings, earning the description synanthropic—"syn" meaning "with" and "anthropic" referring to humans. Common synanthropic rodents are house mice, black rats, and brown rats. The group that includes rats, mice, and gerbils, called Muridae, comprises more than 800 species, about 170 of which live in China.

Only a few murids are synanthropic; most live in the wild, and some of them live in the wildest, most inhospitable places in the world. Gerbils are one group with about 110 species that are adapted to making their home mostly in arid regions of Africa and Asia. Many gerbils survive well in the high, sandy and rocky deserts and grassland steppes of Asia. One species, Mongolian gerbils (*Meriones unguiculatus*), has made human homes its own home, as a popular caged pet. Small, cute, furry Mongolian gerbils have among their cousins a desert-dwelling species, great gerbils (*Rhombomys optimus*, Figure 4.2). Unlike Mongolian gerbils, great gerbils live mostly in central and western Asia, with a stronghold just north of the Tarim Basin (Chapter 2), and one population inhabiting dry washes in the Gobi Desert. Great gerbils are large, as the name suggests—twice the size of pet gerbils—approximately 20 centimeters (~8 inches) long, plus an equally long tail, and weighing about 275 grams (over half a pound). Great gerbils live in extended families in burrows, with each burrow containing one family, and the gerbils in multiple burrows together form a colony.

Figure 4.2 Great gerbil (*Rhombomys opimus*); Muyunkum desert of South Kazakhstan.
Source: Dmitry Fch/Shutterstock.com

And all this matters how?

Great gerbils, like their relatives black rats, brown (Norway) rats, marmots, jerboas, jirds, and other rodents, are reservoir species for a number of diseases, the sorts of nasty diseases that keep epidemiologists awake at night. Diseases transmitted from animals to humans are called zoonoses (singular, zoonosis). Zoonoses may make up as much as 60% of all human infectious diseases, causing about a billion cases of illness and millions of human deaths yearly. Got your attention yet? A reservoir species, as its name suggests, is one in which a disease-causing pathogen can reside, poised, if you will, to be passed along to humans. Zoonoses include hanta virus, tularemia, Leishmaniasis, Ebola virus, and, the scariest of all, bubonic plague.

THE THIRD PLAGUE PANDEMIC

In 1855, in the far, remote northwest part of Yunnan, that area of rich biodiversity, bubonic plague reared its ugly head. This wasn't the first time that a plague outbreak had occurred. In fact, plague had been known to infect humans since the Bronze Age, at least 3,500 years ago. What was different this time, in what came to be known as the Third Pandemic, was documentation and scientific explanation, creating a framework to understand the disease and how it came to infect humans. And this explains why we begin with the Third Pandemic before reviewing the plagues of ancient history. Plague is really a disease of wildlife, with occasional spill-overs that affect humans, as the zoonotic cycle of the disease. But a pandemic is more than simply a "spill-over." Recognizing the context of the disease and wildlife, let us continue the story.

That same biologically rich area of Yunnan is equally diverse in minerals, especially copper. Large numbers of Han Chinese had flooded the area by the 1850s to try to extract their own riches from the extensive deposits. In this instance, the plague re-emerged from its reservoir host, one of the rodents found in that wild area, possibly Asian house rats, *Rattus tanezumi*. As the miners moved around the area and their food scraps and detritus attracted one or more species of rodents, they came in contact with fleas that had been feeding on their Asian house rat host and inadvertently took both fleas and rats with them. They also moved the plague with them.

Over the next few years, the disease spread slowly, first to K'unming, the capital of Yunnan, in 1866, then to the Chinese coast at the Gulf of Tonkin the following year. After reaching the coast, the plague was carried to other coastal cities, arriving in Canton and Hong Kong by 1894. From those two major port cities, the spread accelerated as the plague traveled by sea with maritime trade, reaching Mumbai (formerly Bombay), India, by 1896. Two more years found the plague in Mecca, Madagascar, and Japan. By the dawn of the 20th Century, the plague had reached as far as Portugal, Scotland, and the west coast of the United States. Soon the disease was found on six continents, achieving status as a pandemic.

The term plague, from the Latin *"plaga"* meaning "strike" or "blow" was first used in the 1300s. As a verb, it means to harass, torture, or torment. As a noun, the word is associated loosely with a number of afflictions and major diseases and has been used to describe historical epidemics of smallpox and measles in southern Europe. But the typical use of the word, and the one universally recognized, is to refer to the manifestation of the disease caused by infection from a particular species of bacteria, *Yersinia pestis*.

Often, the term plague is used as shorthand for bubonic plague, which is partly correct. Although there is just one plague, there are three forms of it, whose names describe where in the body the infection occurs. Septicemic plague, the rarest of the three forms, occurs when the bacteria enter the bloodstream directly and multiply, resulting in numerous small blood clots that cause necrosis in tissues, especially in extremities, such as feet and hands. With the body expending all its clotting resources on the small clots, uncontrolled internal bleeding occurs, causing skin discoloration and loss of organ function. Untreated, it is nearly always fatal, and the disease can progress so quickly that a victim may die as soon as or even before symptoms are seen. Scary stuff. The second form, pneumonic plague, is caused by the entry of the bacteria into the lungs, either indirectly from a septicemic infection or directly, through breathing in airborne bacteria from another victim suffering from pneumonic plague. Once the bacteria enter the lungs, the rapid progression of pneumonia leads to death within a few days.

The third form, bubonic plague, is an infection of the lymphatic system. It gets its name from a painful and swollen lymph node, the swollen area being referred to as a bubo, which can grow to the size of a small apple. Bubonic plague progresses more slowly than the other forms, with symptoms presenting 4–10 days after exposure. Most buboes are located in the victim's groin area, a small percentage in the armpits, and only rarely on the neck or other body areas. Bubonic plague is the most common form of the disease, and it occurs when the bacteria enter into the victim's body either through a wound or insect bite and subsequently travel to the lymph nodes. The bacteria can also enter through the digestive system, when the meat of an infected host animal is eaten, or from handling an infected human corpse.

Many recent cases of bubonic plague have arisen through eating infected rodents, such as rats or marmots or, perhaps surprisingly, meat from infected camels. The bubonic form of the plague has lower fatality rates than the other forms of the plague, seldom reaching 90%—as if 90% fatality counts as a "lower" rate! That lower fatality rate is the good news, and bubonic plague can be treated with antibiotics if detected early. The bad news is that the slower rate of progression and lower fatality rate of this form of the plague means it is more easily spread over longer distances, which can turn (and has turned) the plague into a pandemic.

Yersinia pestis has no common name other than, perhaps, plague bacteria. Its technical description is a facultatively anaerobic, gram-negative, non-motile, coccobacillus. The technical description may be off-putting, but an explanation of each term may make the name a little easier to understand. Let's define aerobic and anaerobic first. "Aerobic" means living and being active in the presence of oxygen, whereas "anaerobic" is doing the same but in the absence of oxygen. "Facultative" means primarily one way but capable of another. So, "facultatively anaerobic" means the bacteria generate energy (ATP) when oxygen is present, by respiration, but they can switch to generating energy via fermentation when oxygen is absent. "Gram-negative" means the bacteria do not retain the laboratory stain used to classify and identify bacteria (Gram staining). Other infamous, disease-causing, gram-negative bacteria include *Escherichia coli* (gut bacteria famously known by their abbreviation, *E. coli*) and *Pseudomonas aeruginosa* (a common cause of infection after surgeries), both of which have antibiotic-resistant strains. "Non-motile" simply means the bacteria cannot move on their own—*Yersinia pestis* gets around this by utilizing hosts for transport. "Coccobacillus" means the bacteria are long and cylindrical, with rounded ends—in other words, oval-shaped.

DISCOVERY

As with so many scientific advances, the discovery of plague-causing bacteria was made nearly simultaneously by two scientists. In this case, two bacteriologists were working independently in Hong Kong in 1894 to try to uncover the cause of the ongoing plague, which was not yet a pandemic. Alexandre Yersin (Figure 4.3), a Swiss-French physician, was sent to Hong Kong by the Pasteur Institute, while Kitasato Shibasaburo (Figure 4.4), a renowned Japanese bacteriologist, was assigned to the task by the Japanese government.

On arrival, Kitasato was ensconced in the laboratory of the local hospital and given a large staff to assist him. Within two days, he had found bacteria in the blood,

Figure 4.3 Dr. Alexandre Yersin, co-discoverer of the plague bacterium, *Yersinia pestis*.
Source: Rives/Bibliothèque interuniversitaire de santé/Open License

organs, and bubo of a victim and notified the director of the hospital where the work was conducted. Kitasato may have found bacteria, but his observations described them in a way that was vague and likely represented a contaminated culture. The bacteria he described were a Gram-positive streptococcus species that were not found in the lymph glands of victims—where they should have been if they caused plague. He found the wrong bacteria.

Yersin, traveling from Vietnam with a microscope and incubator, arrived just days after Kitasato had begun his study. Unlike Kitasato, Yersin was not given laboratory access or staff, and he had to set up a make-do lab in a hut (Figure 4.5) near the hospital, where he had access to the bodies of plague victims.

Yersin observed a large number of dead rats in the streets of Hong Kong, giving rise to the idea that the rats might be connected to the plague. He inoculated rats with bacteria from a human victim's lymph gland. Within a week, he found that one of the rodents he had inoculated was dead. As he examined it, he found its spleen filled with bacteria that seemed to be the same as in a human victim's lymph gland. His description was clear and precise, the bacteria were Gram-negative coccobacillus, and he was able to culture them. He proposed the connection between rats and humans but didn't know how transmission could occur. Yersin also speculated that the bacteria he discovered in 1894 were the same that caused the Black Death of the 1300s.

Figure 4.4 Dr. Shibasaburo Kitasato, co-discoverer of the plague bacterium.
Source: Wellcome Images/CC BY 4.0

Kitasato's report was first but Yersin's description was more accurate and consistent. The coccobacillus was found in victims' lymph glands and also found in rodents, providing the first glimpse of a route for transmission. Yersin gave the bacteria the original name of *Bacterium pestis*, but fifty years later, researchers learned more about its evolution. Rather than being closely related to other *Bacterium* species, plague bacteria were found to share affinity with 20 or so other species, which were then collected into a single genus, called *Yersinia*, resulting in the new name *Yersinia pestis*, in honor of Yersin. Kitasato is mentioned as a co-discoverer, but Yersin's name is the one remembered, particularly in the scientific name.

The two scientists led remarkable lives, doing scientific work that went well beyond their discovery of the plague bacteria. Kitasato spent six years studying in Berlin and developed, with his colleague Emil von Behring, therapies for use against tetanus, anthrax, and, most important, diphtheria. Any one of those advances, plus co-discovery of plague bacteria, should have earned him fame and notoriety. He was nominated for the Nobel Prize in Physiology or Medicine in 1901 for his work on diphtheria, but the prize went to von Behring—Kitasato was not included. That slight and being forgotten for his work on the plague were cruel fates for a distinguished scientist.

The careers of the two scientists, Yersin and Kitasato, greatly overlapped: in addition to both working on the plague, Yersin co-discovered the diphtheria toxin and

Figure 4.5 Dr. Alexandre Yersin in 1894, in front of his straw hut laboratory in Hong Kong, where he first isolated and described *Pasturella pestis*, now known as *Yersinia pestis*.
Source: HKU-Pasteur Research Centre—Institut Pasteur and Antoine Danchin

Kitasato co-discovered the antiserum to use against the toxin. However, in contrast to Yersin's acclaim and recognition, Kitasato received little credit for his work and is largely forgotten. History isn't always fair.

Alexandre Yersin had joined the newly formed Pasteur Institute in 1889 but left in 1890 and traveled to French Indochina (today's Vietnam) where he enlisted in the colonial health service. It was while he was in that role that the Pasteur Institute in 1894 asked him to travel to Hong Kong. After his discovery of the plague bacteria, Yersin left Hong Kong. In the next two years, he developed a plague antiserum, which he then tested in India in 1897. The trials were mostly unsuccessful and, after disagreements with local authorities in Mumbai, he returned to Vietnam where he founded and directed a medical school in Hanoi. He even founded an agricultural station, where he grew rubber trees and cinchona trees, from which quinine was derived and used against malaria.

MECHANISMS

Yersin had found the plague bacteria in the lymph nodes of victims. For infection to occur, *Y. pestis* has to make its way into a victim's body, either by inhalation into the lungs, ingestion by eating infected meat, or introduction into the lymphatic system or bloodstream through a wound or by a bite from an infected animal. But disease transmission was not well understood in the late 19th Century when the medical community believed that "miasmas" (bad air) were the route of transmission for infectious diseases.

Yersin's replacement for testing the antiserum, made from *Y. pestis*, in India was a fellow Pasteur Institute scientist, Paul-Louis Simond, whose antiserum tests were carried out in the city that today is known as Karachi, Pakistan. Although the results of his tests, like those of Yersin, were mostly unsuccessful, Simond observed that many patients also had small blisters that contained plague bacteria. The pustules reminded Simond of insect bites and he wondered whether their presence might suggest an insect as a vector of the disease. Recalling that Yersin had found plague bacteria in rats, he turned his attention to fleas found on rats.

Just as Yersin had in Hong Kong, Simond found dead rats littering the streets in India. Live rats hosted thick swarms of fleas, suggesting the fleas could be an intermediary in the disease cycle. Denied access to a local laboratory and even the city's hospital, Simond carried out his research in a Karachi hotel. Anyone who has traveled extensively knows the perils of staying in a hotel and having loud guests in neighboring rooms. Imagine, instead, that the person in the neighboring room is quietly carrying out trials with plague bacteria and dissections of plague-killed rats—loud, rude hotel neighbors would seem comparatively pleasant.

As might be imagined, the suggestion that insects might be vectors of the plague bacteria was met with reactions ranging from skepticism at the most generous extreme to outright rejection and ridicule. Part of the difficulty in accepting this idea was that the methods Simond published in 1898 were incomplete, so his tests were not repeatable. Repeatability in science is critical, so that others can check experimental results, and requires great detail about the methods used. Other criticism focused on the small number of tests conducted, a lack of sufficient controls, and the "certainty" of knowledge of that time: rat fleas *did not* bite humans. The number of tests was a legitimate complaint. Experiments that upset the scientific apple cart need to show clearly that new results did not simply happen by chance. Using multiple tests (or replicates) would have given both the experimenter and critic a reason to accept the results. Simond stated that he repeated the test with the same result. With no published details, his statement was difficult to refute, but more difficult to accept.

The lack of a control against which to compare was another flaw. An experimental control allows for testing whether the effect of an experiment is seen only in the experimental group, while no change is seen in a control. In Simond's study, a control would have been difficult because the connection between plague and poverty confounded the experiment. Plague was especially deadly to the poor, who often resided in crowded, unsanitary dwellings, where direct transmission between humans could not be distinguished from transmission by fleas and vermin. What about the "certainty" that rat fleas would not bite humans? As with so many "certainties" about the animal world, especially concerning insects, the claim that rat fleas did not bite humans was wrong. Very wrong.

FLEAS

The common name flea refers to a collection of about 2,500 species with an evolutionary history currently being teased apart. Historically, fleas were classified in the order Siphonaptera, but research just published as this book went into production tells a different story. Using modern techniques, researchers confirmed a hypothesis proposed in 2002 that fleas are a specialized group within the order Mecoptera. Other mecopterans include obscure creatures such as scorpionflies and hangingflies. The 2002 hypothesis suggested fleas were most closely related to snow scorpionflies, which, like fleas, are wingless and jump. The hypothesis that fleas were mecopterans turned out to be right, but it was wrong about which mecopterans they were closely related to. Rather than snow scorpionflies, fleas' true relatives are from the Southern Hemisphere: the Nannochoristidae (no common name exists), whose larvae develop in freshwater streams and adults have wings—very different from fleas. There are two points to remember: (1) fleas did not "come from" nannochoristids, but rather they shared a common ancestor; and (2) that story remains incomplete. So for now, we await further details.

Whatever their ancestry, fleas are small, generally less than 3 millimeters long; wingless; usually flattened laterally (taller than they are wide); with enlarged hind legs for jumping—a typical flea can jump about 150 millimeters (about 6 inches), amounting to 50 times its body length. That's impressive. All fleas are obligate parasites of mammals and birds. Fleas feed with "siphon" mouthparts shaped like a tube, used to pierce a host's body. Some fleas are catholic feeders, with a broad range of hosts. Many others are more specific and use only certain hosts, such the Malacopsyllidae, a family of fleas with just two Argentinian species that primarily feed on armadillos. Despite being fascinating creatures, and some species being important to humans, fleas have not attracted large numbers of scientific specialists to study them. For all their potential economic and health importance, fleas are still relatively little known. Several hundred species are known only from one record from one host, hardly enough to understand host range, variability, or any other aspect of their biology.

Fleas are holometabolous, meaning their life cycle proceeds from egg to larva to pupa and on to the adult stage. Adult fleas find a blood meal, triggering sexual maturation, and females can lay as many as a few thousand eggs during the adult lifespan of a few months. Eggs are laid either on the host or where the host resides, such as in a rodent burrow. Eggs generally hatch within a few days. Unlike the adults, which live on blood meals, flea larvae feed on detritus—dead skin, feces, nonviable eggs—as they pass through three larval stages. The pupal stage, during which the larva is transformed into an adult, is spent in a silken cocoon. Flea pupae are remarkable in their response to the external cues that trigger their emergence as adults; they can respond to cues that indicate the presence of a host, such as increased carbon dioxide from respiration, increased temperature from body heat, or even the vibrations associated with host movement. The pupa, poised to emerge when such cues are present, are referred to as "pharate," and they can remain in that stage for a long time until host cues are noticed.[1]

AND MORE FLEAS

The two greatest flea experts were a British father and daughter. Charles Rothschild, born in 1877, was educated at Harrow in London. Charles (his middle name, by which he was known; his first name was Nathaniel) was a cousin to the Rothschild banking family, whose extended-family wealth in the late 19th Century was the world's greatest private fortune. Charles inherited some of that wealth and worked as a banker, leading the London family bank for five years. His financial acumen and organized, systematic approach served him well in that role. Alas, Charles's true passion was entomology (which we find totally understandable); he was self-taught, and his specialty was fleas. In his short life (he died at age 46), he amassed a collection of more than 260,000 fleas, which included nearly 75% of all the known flea species. Charles's tireless passion included his description of more than 500 species of fleas. At the age of 24 he collected and described one of those species while on a trip to the Sudan and, two years later (1903), he published the description of *Xenopsylla cheopis*, also known as the Oriental rat flea (Figure 4.6).

Charles Rothschild's daughter, Miriam Louisa Rothschild, was born in 1908 and found herself immersed in the natural world, influenced by her father and her zoologist uncle, Lionel Walter Rothschild. She began her life as a biologist by collecting

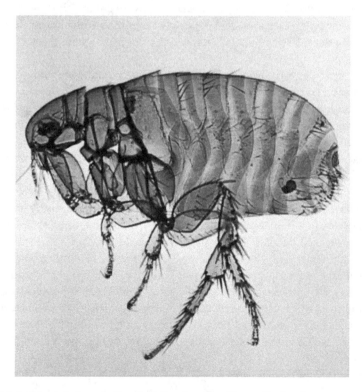

Figure 4.6 Left lateral view of a female Oriental rat flea, *Xenopsylla cheopis*.
Source: World Health Organization, provided by Centers for Disease Control. Image 4633.

insects at the young age of four. Eventually, Miriam, having been schooled at home, longed for a formal education. Soon after the death of her father in 1923, Miriam demanded that she be allowed to attend school, where she threw herself into classes in zoology and literature. Dame Miriam (the title was granted in 2000) was known for her study of all kinds of parasites, and she succeeded her father as the world's leading flea expert.

Dame Miriam figured out how fleas use their hind legs to jump huge distances and, with keen insight and observations, decoded the reproductive cycle of fleas that lived on rabbits, linking flea reproduction to the hormone cycle of their rabbit hosts—an amazing story all of its own.[2] Dame Miriam published prodigiously, becoming an expert in a second field, that of plant toxins sequestered in butterfly caterpillars, which, coupled with warning coloration, confers protection from predators. For someone lacking a formal advanced degree, she held her own and rubbed shoulders with many of the great biologists of the 20th Century, and was named a Fellow of the Royal Society in 1985. She and her brother Victor were the first sister and brother to become Royal Society Fellows.

If Americans are aware of fleas, it is because the insects are pests of their pets. Cat fleas (*Ctenocephalides felis*) and dog fleas (*Ctenocephalides canis*) are the species most commonly encountered. Interestingly, most fleas found on dogs are cat fleas, and dog fleas can also be found on cats. Regardless, both are pests of domestic animals, and both cat and dog fleas are big business. Roughly $9 billion (USD) is spent annually on flea control, treatments, and prescriptions in the United States. In addition to alleviating the suffering of pets, we spend that money because fleas also bite the pets' owners.

Like cat fleas, dog fleas, and rat fleas, human fleas can transmit zoonotic diseases (see Chapter 6), and fleas are found worldwide with a countless list of mammal hosts. One flea that may have escaped the average person's notice is *Tunga penetrans*, which at 1 mm long is the smallest flea (unengorged) but causes huge problems for humans. This flea defies the rule that fleas are ectoparasites but follows the rule that one never should say "all" in describing any insect. *Tunga penetrans*, which has more than a dozen colloquial names, can be found at or just below the surface of the ground, especially in sandy areas. Unlike other members of the genus, *T. penetrans* has a wide host range and is found worldwide, having been distributed inadvertently by humans. Its first description was from symptoms seen on the crew of Columbus's ship, the Santa Maria, shipwrecked on Haiti. When a female *T. penetrans* comes in contact with a host—usually a bare foot—the flea burrows into the skin, leaving only an orifice at the tip of its abdomen exposed, allowing for breathing, defecating, and mating. Free-living males mate with the embedded female, then die. Females produce as many as a thousand eggs, which are expelled and fall to the soil surface, after which she dies. Because she is embedded in the skin, the orifice is an entry point for bacteria in the environment, which can cause serious infections. Mighty big effects from the world's smallest flea.

PLAGUE, AGAIN

Paul-Louis Simond had suggested a mechanism for the transmission of bacteria from rat host to rat flea to human victim, but he stopped short of explaining how bacteria entered a victim's body. That fact, too, lessened the acceptance of Simond's findings of plague-carrying fleas. It wasn't until 1914 that the mode of entry was understood.

Two British scientists, Arthur W. Bacot and Charles J. Martin of the Lister Institute of Preventive Medicine (where Martin was director), studied the Oriental rat flea, *Xenopsylla cheopis* and found that when a flea fed on a plague-infected host, bacteria (*Yersinia pestis*) formed a "plug" in the flea's foregut, blocking passage of a blood meal. The flea, which was not receiving any nutrition and whose gut was not increasing in size because of the plug, would move to a new host in an effort to get a meal. At each subsequent feeding attempt, the flea would insert its mouthpart and regurgitate, which, when feeding on a new host, spread *Yersinia pestis* from the plug in its gut. As a result, Oriental rat fleas, whose cosmopolitan distribution surpassed their name, seemed to be the ideal vectors of *Y. pestis*. With the discovery of *Yersinia pestis* as the causative agent for plague, the description of Oriental rat fleas by Rothschild, and the feeding plug described by Bacot and Martin, combined with Simond's pronouncement of brown rats as host of the fleas and disease, the explanation for the ongoing third pandemic was complete.

While the ongoing studies contributed to the understanding of the factors underlying the plague, the disease had bloomed into a pandemic, affecting all inhabited continents. Movement across oceans along maritime routes involved black rats (*Rattus rattus*) as reservoir species for fleas and plague bacteria. The first two pandemics, occurring many hundreds of years earlier (Chapter 5), mostly affected Europe and the Mediterranean region. The third pandemic had its primary impacts in South and Southeast Asia.

Starting in China, the plague affected that country and Southeast Asia, with perhaps close to 2 million deaths, most of which were in China. Hong Kong recorded 100,000 deaths in 1894, and that port city exported the disease. A ship that departed Hong Kong in 1896, bound for India, carried plague bacteria in addition to its cargo. When the ship was unloaded in India, plague bacteria were unloaded, too. The plague's arrival was initially not noted but, once established, the disease's effects were vicious. In just over 50 years between the first arrival in India in 1896 and ending in 1948, the death toll from the plague in India exceeded 12.5 million. For context, the world's total death toll was about 15 million; more than 80% of those deaths occurred from 1896 to 1918 (Figure 4.7).

The third pandemic was much shorter than the previous two and was considered to have flamed out completely by 1959. Although this was a plague of India and China, its spread was global. European countries reported about 1,700 cases with 457 fatalities. Counts from South American countries suffered from vastly different counting and reporting processes. That stated, reports of the plague showed Peru with more than 20,000 cases, 10,000 from Ecuador, 7,000 from Argentina, 9,200

Figure 4.7 Interior of a temporary hospital for plague victims, Bombay epidemic, 1896–1897.
Source: Wellcome Images/CC BY 4.0

from Brazil, and fewer than 500 from Venezuela. Data from Africa were even more spotty, and the reports were incomplete. Reported cases from most countries were in the hundreds or few thousands, but more than 20,000 cases each were reported from Egypt and Madagascar and more than 60,000 from Uganda.

Untouched in the first two plague pandemics, Australia was invaded by the third pandemic in January 1900, the first infection being in a Sydney dockworker. Thirty cases were reported within a month, and the comprehensive response of local and national governments kicked in. First, infected individuals and anyone with whom they had contact were quarantined. Through the end of that year, just over 1,700 people were quarantined, of whom 263 were confirmed infected. Second, a thorough cleaning program was begun, to remove trash, manure, and animal bedding, clean houses, and upgrade public sanitation procedures and facilities. Third, a rat eradication program was initiated, resulting in the killing of more than 100,000 rats by government workers, with unknown numbers killed by private individuals. Between the initial outbreak and annual outbreaks over the next decade, the human death toll was limited to just over 500 victims. The fast and thorough response by Australian officials, through the combination of containment, cleaning, and eradication, stopped the potential epidemic in its tracks.

In the United States, just over 500 cases of plague were recorded after its arrival in San Francisco in 1900, of which 64% were fatal. However, unlike the situation in Australia, plague became established throughout much of the western United States, due to a lack of a coordinated response. Two cities where the eradication effort was successful were the Gulf of Mexico ports of Galveston, Texas, and New Orleans, Louisiana. The arrival of the plague in San Francisco and Los Angeles, California, resulted in the bacteria becoming a permanent resident, due to a set of interacting and favorable environmental and ecological factors. The ecology and spread of the plague have been likened to that of an invasive plant: establishment, quiet spread, then flare-ups in distant areas.[3] Its establishment in the United States was helped by the presence of susceptible reservoir hosts and fleas capable of transmitting the pathogen. Rodents serving as reservoir hosts included ground squirrels (e.g., prairie dogs and marmots) infested by fleas commonly found on rodents (e.g., *Oropsylla*) and human fleas (*Pulex irritans*). Those hosts and vectors still actively infect humans in the western United States, more than a hundred years after arrival.

EXPLANATIONS, MISSED OPPORTUNITIES

There are a number of ironies in the story of the third plague pandemic, plus a few remaining questions. Far better ability existed to record events and numbers of cases or fatalities than for the preceding pandemics hundreds of years earlier. Despite the better capacity to chronicle the pandemic in detail, the opportunity was missed. Reports were sketchy, methods for collecting information differed, not only among political entities, and the methods also changed while information was being collected.

Examples of missed opportunities include World Health Organization (WHO) reports, which should display definitive means of collecting, collating, and distrib uting numbers of cases or fatalities by country, state or province, or year. A report can be only as good as its weakest link—in this case, varying completeness in the data, or inclusion of micro-level details but without a macro-level perspective. As an example, the WHO report on the onset and spread of the plague provided details about cases and fatalities in small communities from the 1930s to the 1950s but missed larger patterns at the beginning of the epidemic—when most of the fatalities occurred—or failed to provide countrywide totals of cases and fatalities. But, it is easy to second-guess data collection and display of results after the fact; far more difficult is planning for robust data collection and categorizing while the situation itself is unfolding.

Medical scientists searching for factors to help understand the plague's genesis and spread were quick to come to a conclusion in a series of nearly death-defying experiments—finding *Yersinia pestis* in victims; finding the pathogen in rats and linking the two findings; looking for and finding fleas, discovering the blockage of the flea's foregut, then concluding Oriental rat fleas were the vectors. Consider how the scientists could function without face masks, nitrile gloves, or other modern means of avoiding catching or carrying the disease. The model that was developed for the Third Pandemic (rats → rat fleas → humans) seemed to fit well. That model was then assumed to explain infection routes for previous pandemics. Chapter 6 will tease apart those assumptions and discuss how they influenced interpretations to the detriment of understanding the pandemics.

One aspect of the third pandemic that modern-day information captured was the spread of *Yersinia pestis*. The bacteria that first infected people in the far reaches of the Yunnan province took a year to reach the province's capital city, another year to reach a coastal town, then more than 25 years to move along the coast to Hong Kong and Canton. Add two more years to reach India, then two more years to reach the rest of the world. The plague had reached 77 of the world's port cities by the year 1900. The rate of spread within China was fairly slow, comparable to the earlier movement of caravans along the Silk Roads. In contrast, the spread worldwide moved at the pace of newly minted steamships, perhaps one of the first examples of the world being tightly linked temporally.

What can the third pandemic tell us about future outbreaks and plagues? Far less than we might hope. Although *Yersinia pestis* will still be the main player, there are countless strains of this species. One recent study examined fifty-six strains just from the Tien Shan and other mountains along the Silk Road in Kyrgyzstan. Fifty-six strains, just from a limited mountain region of one country? That alone does not bode well for future predictions. Also, the way that *Y. pestis* changes in infectivity makes it difficult to make any predictions about its future ability to cause an outbreak. The host mammals? We know that a large number of rodents are potential reservoirs for plague bacteria, differing across the various regions of the world: from gerbils in Central Asia to ground squirrels in the American West. What about the vectors of the plague? Aha. That discussion, as well as comparisons to older pandemics, belongs in Chapter 6.

NOTES

1. While in graduate school, RNW and family moved into a house that had been vacant for several months. The previous tenants had owned dogs that spent at least part of their time indoors. Almost immediately after moving into the house, large numbers of adult fleas emerged and began biting, though the biting was not equally distributed among family members. Experiencing first-hand this phenomenon and the differential biting pattern was fascinating, though far less fascinating and far more disconcerting for a spouse and young son, both of whom received the disproportionate share of flea bites.

2. Dame Miriam Rothschild was studying rabbit fleas (*Spilopsyllus cuniculi*) when she discovered that an adult female flea could detect changes in the levels of hormones in the blood of a pregnant rabbit as the rabbit neared birthing young. The changed levels of two steroids caused sexual maturation in the flea, and the female flea then moved onto the newborn rabbits to feed and lay eggs. After 10–14 days, the adult female fleas returned to the mother rabbit and remained there until the cycle was repeated with subsequent rabbit births.

3. The analogy of the bacteria as an invasive plant came from S. D. Jones et al., 2019, "Living with Plague: Lessons from the Soviet Union's Antiplague System," *Proceedings of the National Academy of Sciences USA* 116 (2019) on the Soviet Union's anti-plague efforts.

CHAPTER 5

ᕬᕰᕬ

Not Just the Plague

History teachers should be lionized. Seriously. Teaching history, especially ancient history, to marginally interested students is a formidable task, and teaching it in a way that catches the interests and imaginations of students is a talent that needs to be recognized and honored. The Middle Ages is a subject likely to evince boredom and eye rolling from students who cannot imagine a less interesting subject. One of us, who will remain nameless to protect the guilty, was one of those students. The Middle Ages—to the one of us who barely missed being born and living in the Middle Ages—had to be the dullest and least interesting period in the history of history.

Guess what? Totally wrong. They may have been the *most* active years and thus among the most interesting. The Middle Ages represented a thousand-year period from the end of the 5th Century to the close of the 15th Century, roughly between the demise of the Western Roman Empire and the initiation of the Age of Discovery. Tales of the Middle Ages that made their way into most history classrooms in the United States focused on what transpired in Europe, but plenty happened everywhere else in the world. And that stuff is interesting! Like what? Let's consider just the early part of the Middle Ages, say 500–700 CE (from here on, dates are assumed to be the Common Era unless otherwise specified). In the Americas, the Moche and Nazca cultures along the Pacific coast of South America had reached their respective cultural peaks, only to vanish by the end of the 8th Century. Near the mouth of the Amazon River, an extensive chiefdom and culture emerged, centered on large, earthen ceremonial mounds and trading more than a hundred kilometers in all directions.

Further north, in Mexico, Teotihuacan, the large city-state just outside modern Mexico City, was home to more than 100,000 people and the still-impressive Pyramids of the Moon and Sun. At the time, Teotihuacan was the largest city in the Americas and among the largest cities in the world. Still in Mexico and extending into Central America, the Mayan city-states reached their acme with more than 60 kingdoms, in places such as Tikal, Copán, Chichén Itzá, and Calakmul. The vast complex of Calakmul,[1] at the time the most powerful kingdom in the Mayan

The Silken Thread. Robert N. Wiedenmann and J. Ray Fisher, Oxford University Press. © Oxford University Press 2021.
DOI: 10.1093/oso/9780197555583.003.0005

world, had more than 50,000 residents and two of the three largest pyramids in the Mayan realm.

East of the prime meridian, plenty was happening. In North Africa, the powerful Christian kingdom of Aksum dominated trade from the Mediterranean to the south. In eastern and southern Africa, development of the Swahili culture and languages resulted in wealthy city-states in what is recognized today as Zimbabwe. Asia, with its already ancient cities and advanced cultures stretching from Japan and China in the east to India in the south, was connected to western Asia and Europe via the Silk Roads. Knowledge in the Islamic world of western and southwestern Asia was thriving, with major advancements in mathematics (e.g., algebra was developed and the decimal system advanced), chemistry, astronomy, and medicine (e.g., accurately describing the human circulatory system).

So, cultures, trade, languages, science, and history were lively and plentiful throughout the world in those active years of the 6th and 7th Centuries; history is not just as recorded in Europe. Maybe linking together all that happened simultaneously across the globe might have kept the attention of that bored schoolboy. To help link together some of that history, we offer a historical tale woven from the silken thread that began with silkworms in Asia, all the while acknowledging up front our Eurocentric bias.

THE DARK AGES

The active period known as the Middle Ages is often subdivided into the Early (about 500–1000), High (about 1000–1250), and Late (1250–1500) Middle Ages. It is in the Early Middle Ages, once referred to as the Dark Ages, that this story begins. After the reign of Roman Emperor Romulus Augustulus ended in 476, the Western Roman Empire was all but dissolved, leaving the Eastern Roman (Byzantine) Empire as the only remaining part of that once-vast empire.

The term Dark Ages was often used in reference to the changes that occurred in the aftermath of the fall of the Western Roman Empire. At its peak—and including the break-away Gallic and Palmyrene Empires—the Roman Empire encircled the Mediterranean Sea and controlled Europe as far north as present-day Netherlands, Belgium, and Great Britain. In the east, the empire included much of the land surrounding the Black Sea and abutted Persia and its territories.

The consequences of the Roman Empire's demise included loss of economic power gained through manufacturing and trade, as well as the diminution of well-developed cultural practices in the arts, languages, and literature. Overblown or made-up tales of debauchery and violence are associated with the Roman Empire, but many Roman inventions or advances are still used today. For example, the language of the empire, Latin, was the precursor of the Romance languages (including, French, Italian, Spanish, Portuguese, and Romanian) spoken today across much of the world. That time period is also associated with migrations of people throughout Europe, though whether the migrations were a cause or consequence of the empire's decline is debatable. Those societal changes provide one explanation of the term Dark Ages—cultural, political, and economic losses, coupled with the loss of much of recorded history, which drew a curtain across centuries of European life.

Another origin for the descriptor Dark Ages was more far more literal. Skies in much of Europe and Asia were darkened, beginning in 536 and lasting for more than a decade. That darkening is now believed to be due to two or three volcanic eruptions that spewed ash into the atmosphere, blocking sunlight and leaving large swaths of the Northern Hemisphere in semi-darkness. A geological study published in 2018 analyzed ice cores from a glacier in the Alps bordering Switzerland and Italy, and the results offer insight into the early historical eras. The first of the eruptions, in 536, was of an unknown volcano, almost certainly in Iceland, based on comparing chemicals in the ice core with the composition of rocks there. A second volcanic eruption followed in 540, but this eruption was in the tropics, possibly in Indonesia or Central America. Another layer of ash, dated to 547, was found in the ice core, but that eruption remains a mystery.

One consequence of the sequential eruptions was that average temperatures in high latitudes of the Northern Hemisphere decreased by about 2°C. Two degrees? So what? That sort of drop in average temperatures may seem insignificant, but the cooler period caused major environmental changes. Tree-ring data from around the world, showing slowed growth, indicate the global effects. Crops failed across Northern Europe and Asia. Summer snows fell in China in 536 and for the next two years, causing severe famine that killed a large percentage of the Chinese population. Scandinavian countries faced some of the worst effects. Starvation hit much of the northern region. Swedish villages were abandoned, with some accounts citing as many as 75% of Swedish villages left empty. The unrest and migration from Scandinavia are thought to have led to the later increase in the number of Scandinavian explorers setting out on voyages of discovery. In this telling, it was the turning of bright day into darkness across Europe and Asia that created the Dark Ages. Either explanation—volcanic eruptions or the cultural void following the loss of the Western Roman Empire—would suffice. But another carries even greater import and involves and is defined by insects.

JUSTINIAN AND THE BYZANTINE EMPIRE

The Western Roman Empire fell after its defeat by the Heruli forces under King Odoacer at the Battle of Ravenna (modern Italy) in 476, but the Eastern Roman Empire survived. Also known as the Byzantine Empire, it was centered in the city of Constantinople, known today as Istanbul, Turkey. Early in the 6th Century, Constantinople, named for the early Emperor Constantine the Great, emerged as a great city during the rule of another emperor, Justinian, who ruled from 527 until 565. The Byzantine capital city's population grew to an estimated 500,000 people, the largest in the world at that time, and it was a bustling port along the Bosporus, connecting western Asia's Black Sea with the Sea of Marmara, itself connected to the eastern end of the Mediterranean. Constantinople was also where European and

Asian land trade routes converged—the western terminus of several branches and extensions of the Silk Road.

Emperor Justinian ruled an empire embroiled in wars, riddled with corrupt government officials, and saddled with high taxes, all of which enraged the populace. On the positive side, Justinian was known for simplifying the Roman laws, and many of his changes have been modeled by legal codes around the world. Interestingly, several laws he enacted provided protection to women at a time when that was highly unusual, including economic and legal protections, safety measures, and the protection of prostitutes from exploitation and abuse. Remember, this was 1,500 years ago. However, the rewritten laws also eliminated loopholes used by the upper class to bend the legal system in their favor, thus the aristocracy did not support the emperor or his new laws.

All of these simmering factors came together in 532 in the form of a major uprising, known as the Nika riots, which resulted in great fires that destroyed much of the Byzantine city, including the Hagia Sophia, the Orthodox Church's primary cathedral. Justinian considered fleeing, but his wife Theodora announced that she was staying, convincing Justinian to remain. He stayed and barely retained power through a combination of negotiation and appeasement. Oh, and he also ordered his loyal army to slaughter the thousands of remaining opponents. The outcome was that he retained rule of the empire for another 30 years. Through the remaining decades of his rule, he attempted to re-establish the strength and glory of the Byzantine Empire, largely by reconquering some of the empire's previous territories around the Mediterranean.

Justinian was also known for great building projects, such as rebuilding the Hagia Sophia. Although, the rebuilt church (actually the third one on that site) bore little resemblance to its predecessors in shape and size; it was, at the time, the largest cathedral in the world.[2] With its high dome and exterior walls clad in marble, it would have been a dominant, awe-inspiring structure, a sight to behold when approaching the city from afar. Inside, the cathedral had more than 100 structural columns, the largest being more than 10 meters tall. The Hagia Sophia, rebuilt through the efforts of more than 10,000 workers, was completed in less than six years. By way of comparison, the slightly larger Notre Dame Cathedral in Paris required 100 years to build.

The great city of Constantinople did not escape the famine that followed the dimming of sunlight caused by the volcanic activity of 536 and after. Feeding the empire's large populace required a large and complex food supply chain. However, all the agricultural areas that fed the Byzantine capital had suffered multiple crop failures, and so the city's starving residents weakened, and many thousands died. The factors that made the city great—its position as a trade center with big markets at the confluence of maritime and land trading routes with a concentration of people in a small area—also spelled doom for the city's residents, but not because of the famine.

Just over a decade after Justinian became emperor, the city found itself besieged by an enemy that was more dreaded and more savage than the Huns of southern Russia and eastern Europe or the Sasanian Empire to the east. Already starved, the city's residents were more susceptible to a new enemy from the East—the plague. The period of the Middle Ages, bounded by the fall of the Western Roman Empire and the Age of Discovery, was also bounded at both ends by the first and second pandemics of bubonic plague, the first one beginning in 540.

THE FIRST PANDEMIC

The first pandemic was known as the Justinianic Plague, given its occurrence during the reign of Justinian and its effects on Constantinople and the Byzantine Empire. When the first plague pandemic landed in the Byzantine capital, it landed hard. The plague's origin has been debated for decades. At one time, the bacteria (*Yersinia pestis*) that caused the plague was pegged as coming from Ethiopia because the first signs of the plague appeared in Egypt in 540. From Egypt, the plague moved eastward along the Mediterranean coast, then sea transport brought the disease to Constantinople in late 541 or early 542. However, just because the plague was first recognized in Egypt doesn't mean it originated in Africa. Modern research using molecular methods point instead to Central Asia as its origin. Movement of the pathogen from Asia—traveling first to Africa and then to Constantinople—occurred along the Silk Roads and its maritime extensions.

Bubonic plague gets its name from the appearance of growths on the body, termed buboes. Once in the lymphatic system of a victim's body, the bacteria are carried to lymph nodes, typically those located in the neck, armpit, or groin. The buboes swell as bacteria reproduce and accumulate and, in late stages of the disease, can be the size of a small apple. Bacterial invasion of organs, such as the liver or spleen, hastens the victim's death. Plague symptoms, such as fever, headache, and nausea, may first appear within 2–6 days after infection, but disease progression from that point is swift and severe. Invasion of the lungs as a secondary infection not only spells the impending death of the victim, but the infection can then be spread rapidly throughout a population as bacteria are passed through the air by coughing; inhalation by others can then cause pneumonic plague in new victims. Mortality for bubonic plague—estimated using modern strains of the bacteria—ranges between 40% and 60%; for pneumonic plague, over 90%. Plague is not something anyone would want to experience. Perhaps the only positive thing about it is that death follows soon after infection.

Once the plague appeared in Constantinople, the disease quickly affected the city and its residents and with frightening consequences. Still, not everyone infected by the plague died from it. Fittingly, Justinian's Plague infected Emperor Justinian, but he was one of the fortunate who recovered. The death toll of the less fortunate was enormous. Estimates of city residents dying from the plague appeared in various personal journals of the time. Court historian Procopius wrote of estimates of 10,000 dying *daily* (other reports are in the range of 5,000 daily deaths) and his writings told of the large numbers of accumulating corpses, because there was either no one to bury the dead or no one willing to come in contact with the dead. Procopius wrote of enormous pits that were dug to hold 70,000 corpses, but with 5,000–10,000 dying daily, such pits would be filled quickly, and many pits would have been needed. Because of prevailing church doctrine, cremation of bodies was not allowed. And so, bodies were stacked in churches and wall towers, accumulating. The corpses did not improve with age.

The true death toll would have been difficult to derive accurately, even by those living through the plague. Counting the numbers of people dying of disease was compounded by the famine caused by the 536 volcanic eruption. Some modern analyses suggest that Constantinople lost between 33% and 50% of its population, with as many as 10 million deaths in western Asia and southeastern Europe over the eight-year period 541–549. Estimated numbers of deaths for the pandemic overall are all across the board. Many of the guesses converge on about 30–50 million deaths across the geographic and temporal extent of the pandemic, but some estimates are twice that number. At the other extreme are claims that because the "maximalist" estimates are not supported by either writings or physical evidence they have no basis, and further, the plague caused far fewer deaths and had minimal lasting effect. Given that there was no official historical accounting made, it is difficult to accept completely the numbers of fatalities reported. Whatever the true numbers, death from the plague would not have been pleasant in the mid 6th Century.

THE PANDEMIC CONTINUES

The worst of the epidemic in Constantinople abated within roughly four months, though flare-ups in the city and nearby countryside occurred for another 50 years. The fear inspired by the plague must have been almost as debilitating as the disease itself. Explanations for disease were consistent with other knowledge at the time—very limited and influenced by misinformation. Diseases were caused by miasmas (bad air), appearance of comets or the position of stars, arrival or presence of foreigners, poisoning, or, especially, the wrath of an angry deity.

Although the plague's daily death toll dropped dramatically after a few months, the plague's consequences were felt for many years. Compounding the effects of the ongoing famine, the loss of as much as 40% of the population by disease meant a weakened military, as fewer new soldiers were recruited, and fewer young men were even available to recruit. In addition, the tax base needed to pay for military or other Justinianic projects was sharply reduced. The loss of agriculturists only worsened the effects of the famine because there were fewer survivors who were able to grow the food needed by the city's populace. Constantinople went on, but badly crippled.

Apart from the goings-on in Constantinople, the Byzantine Empire continued its military campaigns and trade, both of which potentially influenced the spread of the plague. Movement of soldiers and supplies for battles also moved the disease. The ongoing Gothic War took place in Italy, Sicily, Corsica, and Sardinia—where, perhaps not coincidentally, major outbreaks of the plague occurred. The Lazic War against the rival Sasanian Empire took Byzantine resources and the plague to Persia. All the losses—military, economic, population—led to a weakening of the empire. The empire went on, but it too, was badly crippled.

All around the Mediterranean, outbreaks of the plague flared. Some outbreaks flamed out quickly, others smoldered locally for long periods. The plague also spread inland, the speed of its forward progress defying explanation, as it made its way throughout much of the European continent as far north as France and southern Germany, as well as England and Ireland. Wherever the plague arrived, fear and misinformation about its cause also spread, just as in Constantinople. The disease's incursions into all domestic life were worsened by the lack of any understanding of how it spread or any ability to form a rational response to it.

The propensity for humans to want to understand phenomena, to accept misinformation, and to point a finger at whomever is blamed seems to have been as strong in 600 as it is in contemporary life. As the disease spread into Germany, one explanation was that water supplies had been poisoned by Jews. The resulting persecution and massacre of Jews during the pandemic was so horrible that the Pope issued an edict prohibiting any further violence against Jews and their property.

The worst of the plague ended within a few years in Constantinople but, overall, the pandemic persisted for about 200 years. Its spread was fast and its effects were harsh. Still, at this bookend of the Middle Ages, no one knew where the plague came from or how it spread. It would be another 1,500 years before light was shed on what the plague is and how it infects humans.

THE BLACK DEATH APPEARS

The other bookend of the Middle Ages sits between the 1300s and 1500s, delineated by the onset of the Age of Discovery. This was a time of explorations, undertaken initially by the Portuguese, such as the voyages of de Gama, Cabral, and two explorers named Dias—first Dinis, later Bartolomeu—when the patron of voyages, Portuguese Prince Henry the Navigator, recognized their value and supported them.

Later in the Middle Ages, more explorers from other empires added to the oceanic voyages, mostly seeking a route to the lands of the Orient, with their perceived vast riches and exotic products. Some of those later explorers included Cabot of England, Cartier of France, and Spaniards aplenty, such as Balboa, de Leon, and Magellan (although Magellan was Portuguese by birth, he sailed under the Spanish flag). Columbus exemplified the connections among empires seeking the passage to the Indies and Spice Islands—Italian by birth, later based in Lisbon and sailing for the Portuguese merchant marine, and then funded by the Spanish court of Ferdinand and Isabella. The enslavement and early settlements in the Americas resulting from his voyages figure prominently in Chapters 9–11 on sugar and yellow fever.

The onset of the Age of Discovery coincided with the conclusion of the far-from-dull Middle Ages. It turned out that quite a bit happened over those 900 years. Another way to mark the latter bookend of the Middle Ages was by the arrival of another pandemic of bubonic plague. The second pandemic began as an outbreak, first in Mongolia and northeast China in 1331, and was followed by the plague's spread westward and the arrival of the Black Death in Europe and Western Asia, beginning in 1347.

It was only much later that the term Black Death came into use. During the pandemic, the plague was known as the Pestilence. The term Black Death did not, as one might guess, refer to the color of extremities in which gangrene had taken hold. Most likely its derivation is from the Latin *atra mors*: "*atra*" means "dreadful" or "black," as a mood or sentiment could be, and "*mors*" means "death"; *atra mors* then translates to black death.

As with the first pandemic, the most-accepted explanation—created in modern times, not ancient—of the plague's emergence is that it began in the east, in the high deserts of Central Asia, the Taklamakan and the Tarim Basin, along the Silk Road. The protagonist of the story was most likely the great gerbil (*Rhombomys opimus*). These gerbils are large, as the name suggests, being roughly 20 centimeters long with a tail of similar length, and hefty, weighing 250–300 grams. This species, like many other gerbils, is well adapted to arid deserts, living as families in burrows, and with multiple burrows forming colonies.

Other possible candidates for initiating the plague include one or more of the marmots, such as the tarbagan (*Marmota sibirica*) or bobak (*Marmota bobak*). The bobak, native to the steppe, is similar in its ecology to North American prairie dogs (*Cynomys*). Bobaks are a reasonable suspect, as they were considered the reservoir for a plague outbreak in Russia in the late 1800s, nearly the same time that the third pandemic began in southwest China. Gerbils, marmots, prairie dogs, and other rodents serve as reservoirs for a number of zoonotic diseases, the kinds of diseases that are transmitted from animals to humans.[3] The disease in this case was the plague, that ancient killer that is still with us in the 21st Century. To make that point all the more real, consider that in 2020 five residents of the southwestern United States (California, Colorado, and New Mexico) died or were sickened by *Y. pestis*, whose hosts were either ground squirrels (*Spermophilus*), prairie dogs (*Cynomys*), or one of the western chipmunks (*Neotamias*). Also, a 2020 outbreak of plague in Mongolia began when several locals died after feeding on the flesh of infected marmots—either tarbagan (*M. sibirica*) or grey marmot (*M. baibacina*). The plague is still with us today.

Time for a brief pause and a few questions. Plague is a disease of wildlife, especially rodents. Then how do the bacteria get from the rodents to humans? Apparently through fleas, at least initially. Ever since Yersin, Kitasato, and Simond made their discoveries in the late 1800s and early 1900s, fleas and rats have been center stage in the story. As more information about wildlife diseases has accumulated, explanations for disease transmission have been refined. Plague is an active area of research and explanatory models will continually be refined. The general understanding of how disease agents cycle in a host are understood fairly well. The tentative explanation serves as a hypothesis to be tested. Science operates by posing hypotheses, testing them, then refining and posing new hypotheses. Plague has proven to require a complex explanation, and it seems to be far more complex than was originally envisioned. And it is difficult to recreate and test hypotheses from 1,500 years ago.

PATHWAYS FOR THE PLAGUE

The plague didn't suddenly arrive in Constantinople and begin its trail of death there. By the time it arrived in Constantinople in 1341, the plague had already been killing in Asia for a decade, with some estimates of as many as tens of millions of deaths in northeastern China and Mongolia. But again, recall the caveat about estimated fatalities. The disease moved from rodent hosts in the Gobi or Taklamakan deserts of Mongolia and China in multiple directions and by several routes, mostly on threads of the Silk Road: (1) eastward into China; (2) south, where the disease then was loaded onto vessels, for a maritime extension of the Silk Roads on voyages to the Red Sea; and (3) westward, along a southern track of the Silk Road that wound past Samarkand, Bukhara, Merv, and on to Baghdad and beyond.

All of the routes likely contributed to the spread of the disease and the onset of Europe's Black Death. However, the most important route seems, again, to have been westward on one of the tracks that took the disease past the inland Caspian Sea, where the great Volga River empties into it, and on to the Black Sea port city of Caffa on the Crimean Peninsula. In short, the disease spread largely along the various threads of the Silk Road and its extensions.

Caffa (now the Russian city of Feodosia) featured prominently in the mid-1300s, as it became part of the Republic of Genoa, that Italian city-state with holdings throughout the eastern Mediterranean and western Asia. Caffa became the major port for the traders from Genoa, and great wealth accumulated in Genoese hands there.

During the 1340s, Caffa was attacked by Mongols, and a long siege began. The Mongols were already infected by the plague—their home range included the arid land along the Silk Road. According to legend, the Mongols catapulted the bodies of their dead (or dying) troops over the defensive city walls, in what is considered the first case of biological warfare. Supposedly the corpses launched into the city started an epidemic of plague by infecting the residents. Continuing the legend, merchants fled the city by boat, crossed the Black Sea and, on their return to Genoa, stopped at Constantinople. It was the merchants' stop at that powerful city that brought the plague and its devastation back to the Byzantine Empire. The Caffa story and its role in the pandemic have come under closer scrutiny in recent years. Suspicion as to its veracity has been raised because in the 1300s disease transmission was not understood—and would not be for another 500 years. Biological warfare seems both anachronistic and unlikely. Written records allegedly from the time have been shown to be third-hand accounts recorded years after the supposed event occurred. Nevertheless, the story lives on because of the visual imagery.

Modern perspectives of the siege of Caffa have suggested that, perhaps instead of initiating biological warfare, catapulting bodies had a proximal reason. Supposedly, thousands of Mongols were dying daily, leaving the military with no means to bury such large numbers of dead warriors. Some of the dead were dumped into the Black Sea, but the thousands of daily deaths limited that solution. Legend has it that countless corpses were thrown over the city's walls by catapult to create a horrible stench that would, by itself, kill everyone in the city. Or, maybe it would not kill them but make them wish they were dead. Launching corpses then would have been a "sanitary" means of disposal for the Mongols, and the bodies accumulating inside the city's walls became someone else's problem. Sounds good; maybe sounds too good.

Perhaps this is one of those "just so" after-the-fact stories, and the whole catapult tale may be an urban myth. Regardless of the role of catapulted corpses, the plague did, indeed, make its way from the Black Sea to the Byzantine capitol as a waypoint before it was further spread into Europe.

SPREAD IN EUROPE

As with the first pandemic, the devastation wrought by the plague in the second pandemic was extreme. Constantinople was infected, either by ill sailors arriving from Caffa or by ill soldiers who accompanied Emperor John VI into the city after the Byzantine Civil War. However its entry, the plague caused death and destruction yet again in Constantinople, devastation which eventually hastened the empire's defeat by the Ottomans. Ships from Caffa either landed in Constantinople, off-loaded trade goods and passengers and took on more of both for the voyage to Italy and the Mediterranean, or the ships bypassed the Byzantine city and headed directly for the Mediterranean. Those ships carried with them a most unwanted cargo—infected sailors. The ships' arrival in Sicily was met with horror, as most of the sailors and passengers aboard were dead or dying. They were immediately ordered out of the harbor, but it was too late. The plague had arrived.

Everywhere the ships landed, they left behind people infected by *Yersinia pestis* and an outbreak occurred. Outbreaks were recorded in late 1347 throughout the Mediterranean. By the middle of 1348, the plague had arrived in Lyon and Paris, even in London, spread by ships.

Over the next five years, an estimated 25 million people—roughly one-third of Europe's population—died from the Black Death as the disease spread rapidly throughout the continent, making greater inroads and spreading farther than Justinian's plague. Seldom remembered or reported is that a similarly large number of people died in Asia and Africa as were killed in Europe. Uncounted millions died in China. Some Central Asian cities are thought to have incurred as great as 70% mortality and in Western Asia, as many as 30% of Persians died.

Even though the Black Death occurred 800 years after Justinian's plague, there still was no understanding of how the disease was transmitted or how it could be prevented or treated, which led to fear and a spate of false rumors. One rumor was that just looking at the eyes of a sickened person would result in infection. Some doctors refused to see patients or wore grotesque "plague masks" with the breathing holes filled with herbs, which were to keep out the "miasmas" and counteract the stench of death. Some doctors also covered up with a wide-brimmed hat and wore protective clothing consisting of robes made of oil cloth or leather, gloves, and a hood—a 14th Century version of personal protective equipment. "Treating" patients stretches the definition of the word. Doctors took the pulse of a patient by using a stick, and they lanced the infected buboes with knife blades that were several feet long. Not only doctors changed their behavior. Some priests would not administer last rites. Processions of men moved from town to town, beating themselves with leather straps with metal pieces attached at the ends in acts of self-flagellation. And, once more, the blame for the plague was laid at the feet of the Jews, who were again massacred by the thousands.

From those coastal Mediterranean cities, the plague moved inland and did so much more quickly and completely than had the Justinianic plague, reaching as far north as Scandinavia and Scotland. Although the worst of the plague was gone within a matter of five years, the disease was still present for another 400 years.

Something significant that emerged as a result of the second pandemic was the practice of isolating infected—or potentially infected—individuals from the healthy population. Beginning in 1374, port officials in Venice prevented ships carrying infected crew or passengers from entering the port. Three years later, the Adriatic Sea port of Ragusa (modern-day Dubrovnik) established a station away from the city where any travelers who were possibly infected were isolated for thirty days, known as the *trentena*, to determine whether they were infected. Twenty-five years later, Venice set up their own process to isolate travelers from the eastern Mediterranean for forty days, known as *quaranta giorni* or the *quarantena*. From those terms and that process, the modern quarantine is derived. These quarantines suggest that there were suspicions that transmission occurred from infected individuals. Otherwise, the concept of quarantines would have made no sense; containing passengers on ships would have had little effect on miasmas.

TWO PANDEMICS

Two pandemics, separated by nearly a millennium. Well, the actual separation was not quite that great, as remnants of the first pandemic continued to kill for the next 200 years. Were the two pandemics related and caused by the same disease-causing agent? Yes, both pandemics were the result of infections from the same species, *Yersinia pestis*. However, modern analyses of teeth from bodies exhumed from graves have shown that the lineages of the bacteria differed across the two pandemics—that means the bacteria emerged independently from their Asian rodent hosts.

The research finding that the pandemics were from the same species of bacteria, but different strains, answers the question as to whether bacteria remained hidden in reservoir hosts in Europe for the intervening centuries. The answer is no. Instead, the various strains died out, only to be replaced by related, but different, strains from Asia. Modern molecular analysis has shown the relatedness of the strains of *Y. pestis* from the pandemics and also that the pandemics initiated from the same locations in central Asia.

The infections that caused the second pandemic persisted at a lower level, just as they had following the Justinianic plague. Flare-ups in plague outbreaks occurred, some fairly severe. A second epidemic occurred in 1361, called the *pestis secunda*, killing at least 10% of Europe's population, already reduced by the Black Death.

The Great Plague of London was another of those flare-ups. The plague apparently arrived in London on a trading vessel from Holland in 1665 and raced through the city for 18 months, killing as many as 100,000 people (but again, counts vary). With so many deaths in a short period, there was once again a need to collect the bodies of those who died during the night (Figure 5.1). The cry of, "Bring out your dead," is not just a line from a Monty Python film. The Great Plague of London ended about the time of the Great Fire of London (any other 17th Century "Greats" we missed?), and the fire has been proposed as one of the factors that helped end the plague.

One remnant of the Great Plague of London is the link between the nursery rhyme "Ring, a-ring, o' rosies" and the Great Plague: "Ring, a-ring, o' rosies" referring to a rash; "pocket full of posies," the fragrant flowers to keep the miasmas away; "ati-shoo" being the sneeze, an early symptom (in other versions the line is "ashes, ashes," signifying burning victims or their houses); and "we all fall down," symbolizing everyone dying. As great as the story sounds, it is almost certainly not true. Words in the modern, English version came about later and do not align with early versions of the rhyme, and all the explanations for the rhyme seem to have first appeared after World War II. So, nice story, but not true.

Still, there were no answers as to disease prevention or cure. Well, no good answers anyway. One of the most bizarre solutions for prevention was proposed by one of the greatest minds of all time—Sir Isaac Newton.[4] Newton fled London for Cambridge when the plague arrived. He advised staying away from places infected with the plague, but that advice was neither unusual nor helpful. The eyebrow-raising solution he proposed was toad vomit. We are not making this up. Toad vomit. Newton, in a short paper written in 1667, wrote that other antidotes were not nearly as effective as collecting the vomit from a toad that was hanged upside-down for three days in a chimney. The excretions were to be made into lozenges and worn over the affected area (presumably a bubo). Really? Really. Such a recommendation makes promotion of hydroxychloroquine for a 21st-Century virus seem rather blasé. This was the same Isaac Newton who formulated the laws of motion and gravitation,

Figure 5.1 *Plague at Elliant* by John Everett Millais, engraved by Joseph Swain.
Source: John Everett Millais/Public domain

invented calculus, calculated planetary orbits, discovered the prismatic composition of light, and explained supernumerary arcs. And yet, toad vomit? As a world, we should be grateful that Newton chose to focus on mathematics and physics with *Principia Mathematica* and did not instead write *Principia Biologica*.

But how was the bacteria transmitted to humans? The historical answer has been rats, as disease reservoirs and, even directly, as vectors. More recently, the focus has been on the great gerbil, resident of Asian deserts. OK, but how did the disease make its way out of the rodent burrows of Central Asia to, first, Constantinople, and then the entire European continent? It wasn't that gerbils were transported in the diversity of goods carried on caravans across Asia on the Silk Roads. Maybe it wasn't even the gerbils themselves that accompanied the caravans. Instead, maybe traveling on their own were Oriental rat fleas (*Xenopsylla cheopis*) carrying bacteria to new settings, new hot spots—hot spots that had resident rats, either black rats that so often traveled on ships or brown rats, the scourge of cities all over the world. And then, there, the fleas bit humans, causing infections in human victims, after which the fleas returned to rats as reservoir species. And repeat.

Easy. Easy and simple. Just like Newton's toad vomit. Easy and simple, and wrong. In Chapter 6, we explain just how wrong, upending ages of traditional explanation for the disease and the human contempt for rats.

NOTES

1. Calakmul nearly defies description. RNW visited the site (twice) in 2007. At the time, the ruins were not easily accessed, being located some 60 kilometers off the highway. Travel for the first 20 kilometers was on a road that was somewhat paved, though at most one lane of crumbling blacktop, leading to a guardhouse and gate. But for the last 40 kilometers, there was just a one-lane, dirt (mud) track through the forest. After starting along that track the second time, just after dawn, our vehicles encountered both a puma and jaguar about 30 minutes apart. The jaguar gave us a piercing look that stopped us in our tracks, breathless and speechless. On that visit, we arrived at the ruins shortly after dawn, where we were the only visitors. At Calakmul, nearly 7,000 structures are scattered over an area of 20 square kilometers, including two major pyramids, the largest one 45 meters tall—so tall that from its peak we looked *down* onto the canopy and its black howler monkeys. Unbroken forest, 3 million hectares of it spread before us in every direction. Magical visits, magical place. A rich and detailed account of such a visit is contained in an essay written by Barbara Kingsolver, "The Way to Nueva Vida," in the September/October 2003 issue of Sierra Magazine.

2. The Hagia Sophia, 82 meters long, 73 meters wide, is slightly smaller than the Cathedral of Notre Dame in Paris and about 33% bigger than an American football field. Its massive dome rises 55 meters above the ground. It has survived many earthquakes and still stands as one of the true wonders of the world. RNW visited the Hagia Sophia while in Istanbul in 2004, and it is truly a magnificent edifice. Built as a Christian cathedral, it was converted to a mosque after the fall of Constantinople to Sultan Mehmed II, leader of the Ottomans in 1453. Following the independence of Turkey in 1935, the Hagia Sophia became a museum until 2020, when it was converted back for use as a mosque.

3. Diseases of animals that cause infections in humans are termed zoonotic. Most zoo-notic diseases have different cycles: enzootic, epizootic, and zoonotic. The enzootic cycle includes the routine life cycle of the host and vector; in the case of plague, rodent and flea. Great gerbils live as members of extended families connected through complex burrows, which allow plague bacteria to persist without killing every gerbil. The epizootic cycle is akin to an outbreak or epidemic of bacteria, but at a much greater incidence of infection or over a larger spatial scale. For plague, infection by bacteria increases, affecting more hosts and producing more infected fleas. That point is where the zoonotic cycle intersects. Infected fleas are in the proximity of, or in contact with, domesticated animals or even humans, and direct transmission of the bacteria to these new hosts follows. For the flea-rodent-host model, all aspects are affected by external factors, such as temperature, precipitation, and soil characteristics. All of those factors affect availability of food for the hosts, development time and success of the insect vectors, and possible persistence in the soil.

4. When RNW was in London in 1996, the requisite visit to Westminster Abbey included examining where famous persons were entombed. While others spent time looking at the tombs of kings and queens, we focused on two, almost adjacent, burial vaults of two of the greatest thinkers and influential scientists of the past thousand years (though illustrating our Eurocentric bias): Charles Darwin and Isaac Newton. But toad vomit? Really? Maybe Newton was hit on the head by a few too many apples . . . .

CHAPTER 6

✺

Sorting Out the Plague

Television in the United States in the 1970s featured the murder-mystery series *Columbo*, starring actor Peter Falk as the protagonist, Detective Frank Columbo. Columbo was known for his ability to solve murders through his dogged pursuit of the crime from seemingly oblique angles, and his catchphrase, "Oh, just one more thing," was a way to keep the suspect—and the audience at home—off balance and guessing. More recent crime-solving shows, such as the many *CSI* and related shows, use newer forensic technology to solve crimes. A biased writer would say the newer techniques lack the aplomb of a detective with tousled hair and a rumpled raincoat.

Solving television murder mysteries requires answering questions of who, what, where, when, and how. Solutions require sorting among the competing options that arise and demand attention, which can detract from finding out the truth. It is not just eliminating suspects with solid alibis, figuring out who could have been the perpetrator, or how the crime was committed but, equally, who could *not* have been the perpetrator and the many ways the crime could *not* have been committed. And all of this is completed in 55 minutes.

Investigating the plague from ancient history is, in some ways, similar to a good television crime mystery. We know some of the facts. At least, we *think* we do. But do the "facts" survive scrutiny? Unraveling the details and addressing questions to arrive at an explanation of the plague requires the full weight of modern technology and more than a little of Detective Columbo's "Just one more thing."

Presenting the explanations about the plague, its origin, transmission, and effects, would seem to be straightforward. The plague has killed countless millions of people, primarily in three major pandemics, though it is still a killer on the loose today. The duration of each pandemic differed, ranging from just under 100 years to more than 400 years. However, the vast majority of plague deaths in each pandemic occurred in a short period. The disease manifested in three forms—primarily bubonic, but also pneumonic, and septicemic—none of which was very forgiving. The bacterium causing the plague was discovered in 1894 by the bacteriologist, Alexandre Yersin,

The Silken Thread. Robert N. Wiedenmann and J. Ray Fisher, Oxford University Press. © Oxford University Press 2021.
DOI: 10.1093/oso/9780197555583.003.0006

of the Pasteur Institute, nearly simultaneously to the finding of his Japanese counterpart, Kitasato Shibasaburo. The bacterial species that raged in the 6th Century is the same species causing 21st Century outbreaks in equatorial Africa, the western United States, and northern Mongolia.

INVESTIGATING THE PLAGUE

In the most-recent pandemic, which began in the late 19th Century, plague bacteria were vectored by fleas, likely Oriental rat fleas or one of the other rat fleas living on some of the many Asian gerbils and rats. Oriental rat fleas readily move between host species, enabling other fleas to become infected, and in Yunnan, eight species of fleas have been found to be infected by *Y. pestis*. At some point, fleas infected other rodents, most likely Asian house rats (*Rattus tanezumi*) or brown rats (*Rattus norvegicus*; also called Norway rats, although they are not from Norway). Populations of both species have adapted to associate with humans (a condition called synanthropy) and both are hosts of Oriental rat fleas; and Oriental rat fleas transmit plague to humans. The plague slowly reached the eastern coast of China and its port cities, from where the disease was transported by sea in all directions to countries around the world.

The plague needed help to spread by sea. The suspects that fit the description are black rats (*Rattus rattus*), which have been historical inhabitants of ships. As a reservoir host for the bacterium, black rats seemed to be the right choice to explain plague's spread. India suffered the most from the third pandemic. When ships from China docked in India, where people and goods were offloaded, the brown rat became the new reservoir species and the disease ran unchecked through the country, spreading inland at a rate of about 12–15 kilometers per year. The explanation for the pandemic's spread made sense: bacteria infect fleas; fleas infect rodents and humans; then the fleas cycle back to the reservoir rodent host to produce a new generation, since these fleas are unable to reproduce on humans. The cycle was simple, and it allowed for the pathogen to be moved around the world on steam-powered sailing ships, initiating outbreaks worldwide, wherever those ships visited port.

This explanation appears to explain the third pandemic, so it made sense to apply that explanation to the pandemics of the 6th and 14th Centuries. As a result, the tales of rats bringing the plague to the cities of Medieval Europe arose and still prevail. And why not? It was the plague, right? Well, yes and no. It was the plague, but it was not simply *The Plague*, as a monolithic entity. Plague is more complicated than that.

The 19th Century observations and explanations for the third pandemic became the standard, and stories from the first and second pandemics were adjusted to fit the rat-rat-flea model. When 20th Century observations and conclusions conflicted with the standard model, the newer conclusions were ignored, discredited, or discarded. However, recurrence of the plague and new hot spots in the 21st Century, as well as general concerns about zoonotic diseases, have increased scientists' scrutiny of the standard model. Modern methods have been applied to plague research. Previous reports and conclusions have been revisited, spawning new studies to test age-old assumptions and interpretations. The accumulation of re-interpreted evidence, and the robustness of the newer techniques and their analyses, are yielding newer interpretations—just like Columbo building a case.

The story is complicated, but we need to sort the current explanations for the plague from the historical, oversimplified rat-rat-flea model. We will address key points as systematically as we can, to answer the who, what, where, when, and how questions, by addressing:

- the pathogen
- the hosts: as reservoirs and agents of spread
- the vectors
- the origin, timing, and routes to get to Europe
- spread throughout Europe
- other possible vectors
- how and where the pathogen persisted

The list does not align perfectly with the desired who, what, where, etc. But to arrive at the "who," we need to use all the clues possible from both older studies and newer evidence because of the many combinations of species that have been implicated in the great plague mortality. In other words, whodunit?

Exhumation of human remains to solve mysteries sounds like the stuff of television crime or cold-case dramas. What if the cold cases are more than 1,000 years old? For years, archaeologists and anthropologists have unearthed old dwellings, burial sites, and small settlements to look for clues about earlier life. Tools, pottery, and other artifacts can tell us today something about how a civilization's people made their daily living ages ago. Those are hard artifacts.

But what about smaller, seemingly fragile, and invisible artifacts? Like DNA. As an ancient corpse decomposed, most of its associated DNA would have degraded. Most, but not all. Hard artifacts from exhumed bodies include bones and teeth, and those retain DNA. Newer techniques allow the finding of intact DNA from those ancient teeth and bones. Ever see the film *Jurassic Park*? Recreating dinosaurs required retrieving intact DNA from fossils and bringing that DNA to life. Science fiction, right? Perhaps the story was a little far-fetched, but maybe not.

Death records may tell when and how a person died, though the accuracy of the stated cause of death can be questionable, especially as one goes further back in time or if the cause is poorly understood. Burial records tell us the date of a death and are more accurate about *when*, though offering few cues about *how* a death occurred. However, some burial records show clustered deaths from multiple members of a family, suggesting a deadly, swift malady, or other misfortune that swept through a household.

Remains from mass graves or small-town cemeteries with plots from closely clustered dates offer a starting point. Older, local records that told of an epidemic can narrow a search. Added to that, DNA from bacteria that infected a body can be found with newer extraction techniques used for exhumed remains. The stuff of science fiction now can tell quite a bit about what it was that infected and killed the person whose remains are examined hundreds of years later. Evidence left by the killer, the plague bacteria, is retained in the teeth of the victims, and those bacteria have confessions to make.

THE PATHOGEN

The third pandemic spread slowly compared to the rapid advance of disease in the pandemics of the 6th and 14th Centuries. As a result, several studies concluded that earlier pandemics must have been different diseases, such as smallpox or measles, which could race through a population. Those conclusions were well-reasoned—but wrong—and all credible explanations conclude that the pandemics were caused by the plague, not by other possible diseases. The evidence? Teeth. Teeth from archeological sites from those time periods contain not only DNA of the deceased, but also DNA of *Yersinia pestis*.[1]

DNA can also be used to tell the story of how *Yersinia pestis* evolved through history. Just as domestic silk moths evolved from wild silk moths that are still alive today, so too is the ancestor of *Y. pestis* still alive. This ancestor is the soil-dwelling bacterium, *Yersinia pseudotuberculosis*, which can cause infection in the digestive tract of humans and other animals (e.g., rodents, birds, cats, and dogs). *Yersinia pseudotuberculosis* is not vectored by insects but, instead, is acquired by animals through contact with contaminated soil or plants. We now know that from this stock came *Yersinia pestis*, between 5,500 and 6,000 years ago, during the Bronze Age. At that point, just like its ancestor, *Y. pestis* lived in the soil and could not survive in insects. It took another 2,000 years (3,500 years ago) for an important mutation to arise—a loss of a single gene. That seemingly slight change in DNA enabled *Y. pestis* to live in an insect host and forever alter history.[2]

But the evolutionary history of *Yersinia pestis* can inform us about the three pandemics as well. For example, it is possible that *Y. pestis*, being derived from soil bacteria, is geographically widespread. Perhaps the three pandemics can be explained by local rodents acquiring bacteria from the soil? Or perhaps the bacteria originated in Central Asia, but after spreading elsewhere, they persisted in those new areas and later pandemics are explained by local acquisition of bacteria? Or perhaps our background information is way off, and plague bacteria aren't even from Central Asia! To sort this out, we turn to evolutionary studies.

For species that have expanded beyond their native range (e.g., humans spreading worldwide from Africa; common pigeons spreading worldwide from cliffs around Europe) it can be difficult to determine what their native range actually was. One tool that can help us determine this is genetic diversity. As a rule of thumb, species have a greater genetic diversity *within* their native range than *outside* it. Essentially, this is due to the fact that, regardless of which individuals left the parent population to spread elsewhere, those individuals certainly did not embody the breadth of diversity within the whole species. When genetic tools are applied to the question of the center of origin for *Yersinia pestis*, the results are salient—Central Asia.

The whole story from evolutionary studies on *Yersinia pestis* in the pandemics is complicated, but there are important takeaways. The bacteria that left their homeland in Central Asia and traveled great distances to cause the three pandemics had stories to tell. The first takeaway is that the bacteria causing each pandemic were each other's closest relatives, meaning they came from a single stock population in Central Asia. Each pandemic has its own idiosyncrasies. The bacteria that caused the Justinianic Plague became extinct shortly afterward. Those that caused the second pandemic (Black Death) were all the same strain, but this strain appears to have entered Europe multiple times—multiple entries, but all from one bacterial strain that could be traced back to Central Asia.

THE ROUTES

OK, where are we so far? We have evidence from the teeth of victims confirming the cause of deaths. We have identified bacteria and different strains that caused each of the three pandemics, which tells us that bacteria were not residing at low levels in Europe in reservoir hosts. We know the bacteria originated in Central Asia. But we have not yet addressed how the disease spread or who the hosts were.

The Silk Roads and their extensions carried a vast array of goods in all directions. The major route westward from its Chinese origin reached the Central Asian high desert. From there (Figure 2.2), the major routes of the Silk Road continued westward past the desert outposts of Samarkand, Bukhara, Merv, and on to Baghdad and the lands past the Tigris and Euphrates Rivers. Other routes turned south from the desert, traversed the Hindu Kush and followed the Indus River valley to the Arabian Sea, where maritime routes connected westward to the Persian Gulf and Red Sea. From China, other routes went directly south and further east, connecting Japan, eastern China, south Asia, and the spice islands of the Moluccas (Figure 3.1). Along the primary routes that passed through the high desert, large trains of Bactrian camels (*Camelus bactrianus*) transported prized goods. Those caravans of camels are thought to have carried the plague in all directions. But how? Were rats transported along with the trading goods? Perhaps even the camels themselves helped spread the infection?

The northern route skirted the Tien Shan range, heading northwest, north of the Aral and Caspian Seas. From there, the routes branched, with one branch leading to the Black Sea, the other passing even farther north along what has been called the Fur Roads (Figure 2.2). Could the Fur Roads have been the route for the plague to enter Europe from Russia?

Various theories and their idiosyncratic combinations of vectors, hosts, and pathways offer solutions to the enigma of how the plague entered Europe and began killing in the first two pandemics. The Silk Roads likely figured importantly, especially for the initial movement of *Yersinia pestis* from its ancestral home. It seems certain that the plague came through Constantinople to initiate the first (Justinianic) pandemic, before being moved to Mediterranean coastal cities. The route to Constantinople may have included traders on the major routes of the Silk Roads coming from the southeast, but also those coming from the northeast, the steppe, on the Fur Roads.

What about the route for the second pandemic? The idea that the plague entered Europe from the southeast, through Constantinople, has been largely accepted. But what routes were used to get the plague to Europe? From the southeast, through the Balkans or the Mediterranean countries of Italy or Greece? Or from the northeast, via trade that skirted the northern shores of the Caspian Sea? The modern molecular analyses point to *both* routes likely playing roles. Analyzing the genetic composition of *Y. pestis* allows tracing the routes in reverse, and the results point to a common branching point north of the Caspian Sea, similar to what occurred in the Justinianic plague. From there, one branch headed westward to east-central Europe, while another moved south to the Black Sea, from where ships carried the plague through Constantinople and on to the Mediterranean.

One anomaly about the route of the second pandemic is the occurrence of the plague as it reached western Russia, which conflicts with the theories of the two routes into Europe. Newer explanations for the apparent anomaly suggest a strain of plague bacteria moved *from* western Europe to Russia, not *to* Europe from Russia.

In the third pandemic—the one that affected India the worst—the plague moved along different routes entirely. After residing in rodents in the mountainous southwest of China, bacteria moved along one of the extensions of the Silk Road, as detailed in Chapter 4. What changed an outbreak of plague into a pandemic was the travel by sea, the plague hopping onto sailing ships that were headed to India. Movement on the Maritime Routes (Figure 3.1) set the third pandemic into motion. At least we recognize the routes and understand that the routes for the first two pandemics were different from the third—different, except that they all began from a point in Central and Northern Asia.

MAMMAL HOSTS

The conventional explanation is that *Yersinia pestis* was found in rodents in Central Asia, and bacteria made their way from there to Western Asia and Europe, where rats spread the disease. Rats have been the suspects implicated in tales about the spread of the plague since the start of the 20th Century, and they certainly were important during the third pandemic. However, investigations of the first two pandemics have given pause to interpretations of the part rats played.

The great gerbil (*Rhombomys optimus*) is considered one of the key species in stories about the plague. Its Central Asian home sat astride the routes of the Silk Roads that traversed the high desert, placing the gerbil in proximity to travelers and traders. Great gerbils, like other rodents, are hosts for a number of ectoparasites, including multiple lice and fleas, and a number of pathogens, such as the plague bacterium. Gerbils were likely one of the reservoir hosts for *Yersinia pestis*, but gerbils themselves did not cause plague or transmit the disease directly. Most likely, gerbils were not transported on the Silk Roads, though their parasites likely were.

Climate factors seem to have played a part in the gerbil's story. Volcanic activity almost half a world away darkened the skies and altered atmospheric conditions, lowering temperatures and shifting rainfall patterns. Great gerbils live in colonies, with burrows that can be as deep as 3 meters below the soil surface, insulated from the worst of temperature fluctuations. However, effects from the volcanoes half a world away in 536 CE and the subsequent several years' of darkness apparently also affected gerbils. The dark period that caused the demise of crops also likely affected food sources, leading to large die-offs of gerbils. As large numbers of gerbils died, the remaining fleas and lice that lived on the gerbils concentrated on remaining rodents, then left their hosts, spilling over and seeking other blood meals, which would have put them in contact with other mammals, such as travelers who stopped at caravan-serai and rested on the ground overnight.

Maybe the contact with gerbils occurred via the intersection with the Fur Roads. Fur trading formed its own economic realm, with the furs being used in lieu of money. Animal pelts became fashion items and were used to decorate clothes. Some pelts were highly prized and so there was a differential value established for some of the more precious furs. The numbers of pelts harvested and traded along the Fur Roads must have been staggering and their trade led to great wealth for certain groups on the steppe. A fur not only provided warmth on the cold steppe, but also symbol-ized the wealth of the person wearing it. Fox and several of the weasels (e.g., mink, ermine, sable) were highly prized furs. The magnitude of the value of fur trading kept a regular flow of traders carrying pelts; those pelts also were likely carrying ectoparasites, such as fleas. Adult fleas would have left the pelt once their host died, seeking other blood meals, but the adult fleas would have left behind eggs, buried in or sticking to the fur.

Oh, one more thing . . . among the other species traded as pelts were rats.

In the third pandemic, Asian house rats and brown rats were most likely the rodents that carried the plague to southwest China. Both rats were always in the proximity of infected people slowly making their way to port cities on the coast of the South China Sea. In ports, such as Shanghai and Hong Kong, the plague was carried on board and the role of the Asian house rat or brown rat was suspended for the time being. Detective Columbo moves on to other suspects, but he doesn't forget about brown rats.

Once ships departed from China, black rats became the reservoir species for the bacterium. Oriental rat fleas (or one of its related species) fed on infected rats, then on passengers and crew. In their native habitats in Asia, black rats are good climbers and can reach resources out of the reach of many other rodents. When associated with humans, black rats can access roofs and upper floors, and even the masts and spars of sailing ships. This climbing behavior is reflected in their common names, as they are often called roof rats or ship rats. They tend to live in groups of both sexes and are mostly found in tropical and subtropical regions. The rats seldom move long distances, as they maintain territories where they defend food sources. Clarification: they seldom move long distances on their own, but they can travel (and have traveled) worldwide as stowaways.

Brown rats, although superficially similar to black rats, are quite different. They are native to northern and central Asia, near the Tian Shen mountains and both the Silk Roads and Fur Roads. This is farther north than the range of black rats, and brown rats do indeed have a greater tolerance to the colder temperatures than black rats. Brown rats moved west from Asia as humans moved; they are cosmopolitan in their distribution and not limited to coastal areas. Unlike its arboreal relative, the brown rat stays close to the ground, preferring to dig or swim. Its affinity for water earned it another nickname, sewer rat.

WAS IT THE RATS?

The imagery of plague outbreaks often includes rats. Conflicting reports exist, even from the recent third pandemic, during which both Alexandre Yersin and Paul-Louis Simond reported countless dead and dying rats in their notes. So-called rat falls were reported to have occurred immediately preceding a number of the outbreaks in India and southeast China. But that was the third pandemic. Their reports were from tropical or subtropical port cities. Even during the third pandemic, not every outbreak included sightings of rats, whether dead, sickened, or otherwise. Nor was every rat infected. Sampling of rats in Mumbai yielded about 500,000 rats caught, of which only 3% were infected, hardly aligning with reports of rat falls, with the streets filled with dead rats. And which rat was it? Black rat or brown rat? The reports did not differentiate between species.

And the first two pandemics, the ones in Europe? Written reports at the time seldom, if ever, mention rats, which is strikingly different to reports from the third pandemic. Why not? Were they so common that they merited no mention? One study gives our plague detective a good clue. A study from 2013[3] tells of archaeological digs made at more than a thousand sites in Norway, many of them containing thousands of animal bones. Their study found few rat bones—a total of about 700 rat bones at only 19 of the sites. The site with the largest number was thought to be a rat nest; still, at that site, there were only some 200 rat bones among 25,000 other mammal bones. Essentially no evidence of rats was found at any sites away from the coast, supporting the idea that the few rats that were present had arrived on ships, but did not venture far from the ports. Rats were not present in numbers great enough in Norway, either in port cities or in rural areas, to support the plague and its spread.

Maybe the results in that detailed study were specific only to far northern Scandinavia. Well, actually, no. Other reports from northern Europe, the British Isles, and Iceland failed to mention rats or, if they did, the rats were confined to ports and did not spread from there.

How about southern Europe and around the Mediterranean? The Justinianic Plague was carried to port cities, black rats traveled aboard ships that carried the plague, and outbreaks occurred wherever the ships made port. Rats inhabited docks and nearby buildings in Mediterranean seaports, and so they could have been the initial reservoir for bacteria, first on the ships and then as the ships unloaded cargo and people. Again, black rats were not found away from the docks, so they do not explain inland spread. Excavations at hundreds of ancient archaeological sites around the Mediterranean have unearthed very few rat bones, despite searches that have gone on for more than 100 years.

For both the first and second pandemics, there were no reports of rat falls, when large numbers of dead and dying rats would have been found just before an outbreak in humans. The most likely explanation for the lack of reported dead and dying rats during the first two pandemics is that rats played a very limited role. Black rats may have harbored the pathogen and been hosts for insect vectors, but their role was probably limited to the point of ships reaching and unloading in port towns.

Was the brown rat the culprit that was responsible for spreading the plague? That species is not restricted to ports and is widespread away from the coast. Brown rats are more aggressive and outcompete black rats, supporting the idea that brown rats could have moved the plague inland.

Oh, just one more thing . . . . The first reliable records of brown rats occurring in Europe were at the end of the 17th Century or beginning of the 18th Century. The first two pandemics occurred in the 6th and 14th Centuries.

INSECT VECTORS

What about the insect vectors? Fleas are blood feeders. They have been the suspects for transmitting plague since Paul-Louis Simond found infected fleas feeding on rats in 1898. Most mammals maintain some sort of permanent flea population in either their fur or their nest. Individual rodents, for example, have been found to be infested by as many as 80 fleas and their burrows and nests by more than 100 fleas.

Fleas have mouthparts that act as serrated blades. By muscular movement of the mouthparts, thrusting and contracting, they cut through the host's skin. As the skin is pierced, the flea excretes saliva, containing anticoagulants, so the host's blood does not clot readily. Insertion and movement of the flea's mouthparts causes small hemorrhages beneath the skin and the flea ingests the pooled blood. This sort of extravascular (outside blood vessels) feeding has been observed in fleas infected with *Y. pestis*. Bacteria that are in the flea's saliva are transmitted to the victim's lymphatic system, where the collection of bacteria produces the buboes. If the flea's inserted mouthparts penetrate a blood vessel, feeding is enhanced and *Y. pestis* is regurgitated directly into the host's bloodstream.

Oriental rat fleas (*Xenopsylla cheopis*) and related species (*X. gerbilli*, *X. hirtipes*, and *X. skrjabini*) are cited as ectoparasites of great gerbils, as well as a number of other rodents and mammals. The tale of the interaction of *X. cheopis* with the bacterium has cemented the place of this flea species in the history of the plague. Early reports showed that while feeding on infected hosts, the flea ingests *Yersinia pestis*, and bacteria accumulate in the flea's foregut, blocking passage of blood meals. Later studies have shown that the guts of *X. gerbilli*, *X. hirtipes*, and *X. skrjabini* are also blocked by the bacterium, leading to intensive probing by the flea in seeking a blood meal. The foregut blockage contains more than a million *Y. pestis*, some of which are regurgitated into and infect the host.

Oh, just one more thing . . . . The reports of flea transmission of the plague by Simond and others did not include voucher specimens of the flea—specimens he collected during his studies.

Voucher specimens are needed so that we can be sure the species being studied has been accurately identified. Without voucher specimens, we cannot be certain what flea species Simond reported. But what do we actually know about the number and distribution of species related to Oriental rat fleas? Charles Rothschild described *Xenopsylla cheopis* from the Sudan in East Africa in 1903. It is a real possibility that the fleas from Central and East Asia that initiated the pandemics were other *Xenopsylla*, and the name Oriental rat flea may have been given to a complex of many species, rather than just one. It is difficult to believe that, after all this time, the named vector of a disease as prominent as the plague has not been studied in a manner that allows distinguishing among the dozens of species either described or proposed. This would be akin to Columbo making an arrest, but only later learning that the culprit is one of identical quintuplets, and so the actual identity of his incarcerated culprit is in question.

True zoological parasites do not kill their host, but *Yersinia pestis* did not read the rule book, and so it violates that rule with the vectors *X. cheopis* and other *Xenopsylla*. With no blood meal in the stomach, the flea continues to try to feed, seeking new feeding sites repeatedly, regurgitating bacteria into each new feeding wound. The flea dies within a few days, either of starvation or dehydration, and some even die with their mouthparts inserted in a host, having made a last effort to feed and regurgitating *Y. pestis* in their efforts. Perfect as a suspect that infects its host.

Several days pass before an infection in a host reaches a level that allows a feeding flea to easily ingest bacteria, making transmission easier. Once the host's blood reaches a septicemic level, there is only a narrow window of time before death occurs. If there are many fleas on the host, however, it is highly likely that at least a few fleas have fed and ingested the bacteria. Fleas can and do leave living hosts, even moving among other related host species, but they quickly leave a dead host. Killing a host is important for the transmission of *Y. pestis*, because the infected fleas leave the dead host and seek a new, healthy host, and the cycle continues.

Oh, just one more thing . . . . The fleas have less than five days to effectively transmit the bacteria.

The bacterial plug in the foregut that makes Oriental rat fleas (and the other *Xenopsylla*) great vectors of *Y. pestis* takes 12–16 days to form. When formed, the fleas are effective transmitters of plague bacteria, but not before then. Once the plug is complete, the fleas cannot eat. So, they starve to death in less than five days. Their days as an efficient vector are literally numbered. Maybe Oriental rat fleas were a great suspect for infecting the host, but not for causing the plague's fast spread.

OTHER FLEAS

Maybe other fleas? Northern rat fleas (*Nosopsyllus fasciatus*) also use brown rats as hosts, and they also occasionally feed on house mice, ground squirrels, and even humans. Northern rat fleas are more likely to be found in a host nest than other fleas and they prefer rats that burrow, such as brown rats. All of this seems promising. But brown rats arrived in Europe in the late 1600s or 1700s, after both the first pandemic and the worst of the second pandemic had passed. Although Northern rat fleas can transmit a number of diseases (such as the agents causing Salmonella food poisoning and tularemia), they are ineffective as a vector of plague bacteria.

What about the more obvious choices—the most common human-associated fleas of today—dog fleas (*Ctenocephalides canis*) and cat fleas (*Ctenocephalides felis*)? Both species feed on humans, human-associated animals, and a variety of wild mammals. An extensive survey of flea-infested mammals from all over the world found cat fleas on 20% of them and dog fleas on just 4%. Cat fleas are generalists and were found feeding on a broad range of mammals, including porcupines, mice, foxes, rabbits, opossum, cats, and dogs. By contrast, dog fleas were found only on cats, dogs (and relatives like foxes), and various weasels and martens. Both species feed on domestic cats and dogs (the common names are misnomers in this respect), with cat fleas being the most common. Both species likely originated in Africa, but after centuries of tracking human movements along with their hosts, today they are found worldwide. Importantly, both species were certainly present throughout Europe during all plague pandemics. Stories of cats during the Black Death abound. Some tell of "cat worship" of the felines as rat catchers. Others considered cats companions to witches, and so the cats were killed to curtail witchcraft. The tall tales even included the Pope—allegedly, the Pope issued an edict to eradicate cats—not true, but the falsehood is added evidence of the ubiquity of cats at the time. Similar stories abound for dogs. So far, either species (or both) seem like ideal candidates for plague vectors.

There are two problems. First, although domestic cats were common, they would have needed to be plague-worthy themselves to account for the fast spread of the disease. Unlike Oriental rat fleas, when cat and dog fleas feed on infected hosts, the bacteria cannot form the plug, so the conditions that make Oriental rat fleas so effective at vectoring plague are missing. Cat fleas are ineffective vectors of *Y. pestis*. Just like that, cat and dog fleas are dropped as culprits in the mystery. Even Columbo could not break the cat flea's alibi.

Of the 2,500 flea species, humans encounter three species far more often than all others. We've just discussed—and dismissed—two of them (cat and dog fleas). Now let's consider the final ubiquitous species—human fleas (*Pulex irritans*)—and its potential role in the Black Death. With a common name like human flea, it already seems promising. And they do indeed feed on a variety of animals, including humans and our associates (e.g., cats, dogs, rats, pigs); so that necessary requirement is confirmed. Even more promising, they are likely culprits for the localized, modern plague outbreaks in rural areas of Madagascar and Tanzania.

Well? How about the human flea? Maybe. However, as with cat fleas, the plug does not form in the digestive system of the human flea, so it is not an effective vector. But maybe an efficient vector isn't necessary. *Yersinia pestis* can still survive short periods on the flea's mouthparts. Large numbers of fleas in close proximity to humans would make them possible vectors. Despite the possible survival of the plague on flea mouthparts, as with cat and dog fleas, the speed of the spread does not track with the fleas needing to leave hosts to reproduce. Human fleas, despite their common name, did not originate on humans. All six species of *Pulex* are endemic to South America. The ancestral hosts of *Pulex irritans*, believe it or not, were guinea pigs (*Cavia porcellus*) in the Andes Mountains. South America? If human fleas played a role in the plague pandemics, then how did they get from South America to Europe and Asia in the 6th Century (for the first pandemic) or the 14th Century (the second pandemic)? This is where things get interesting.

Perhaps human fleas got to Europe with their native hosts? Guinea pigs are one of only five species of animals that were domesticated by native peoples in the Americas (the others are turkeys, Muscovy ducks, llamas, and alpacas); they were domesticated for meat over 7,000 years ago. Just as *Bombyx mori* are moths domesticated from an ancestral species of a different name, *Bombyx mandarina*, guinea pigs are domesticated from montane guinea pigs (*Cavia tschudii*) that live in the high Andes. After domestication, they were spread by trade networks well beyond their native range, and today, domestic guinea pigs are found worldwide. But when did they spread from South America? Pedro Álvares Cabral from Portugal was the first European explorer to find South America, landing on the shores of Brazil in 1500. But Brazil was on the opposite side of the continent from montane guinea pigs. This date is also much later than the first two pandemics. Furthermore, DNA evidence suggests guinea pigs themselves arrived in Europe as late as the 17th Century. And yet, human fleas have been found in archeological sites from as early as 12,000 years ago, so human fleas did not use their native host to disperse to Europe.

The current hypothesis is that human fleas used humans and human-associated animals to travel from South America into North America, and eventually crossed the Beringian land bridge—that narrow band of land that was present when sea levels were lower during the postglacial period, about 12,000 years ago. So human fleas were likely in Europe at the time of the pandemics. However, it turns out this still isn't enough. Remember, plague bacteria in human fleas are not able to make the plug that renders Oriental rat fleas such efficient vectors. Without that plug, human fleas are not able to transmit plague at a rate that is consistent with its spread during the first two pandemics—the spread just happened too fast. What about modern cases where human fleas *are* implicated as vectors for plague outbreaks? Well, the modern outbreaks invariably occur in smaller, often isolated, regions and without rapid spread. In short, fleas that do not make the plug (e.g., cat, dog, and human fleas) are not effective vectors of widespread plague, and that lack of effectiveness rules them out as culprits for the first two pandemics.

We are not totally dismissing fleas. They certainly played a role, and they were *the* vectors of *Y. pestis* in the third pandemic. In the first and second pandemics, Oriental rat fleas indeed brought the plague from the Central Asian deserts to eastern Europe. But perhaps that alone was their major role. Infected fleas can be transported for long distances or periods without feeding, so they could have traveled in clothing or other traded goods. However, it seems they were not adequate culprits for the spread away from the Mediterranean coast, into Europe. Two factors matter for that rapid inland spread throughout the continent: the need for a reservoir species, as these fleas did not reproduce when they fed on humans; and the need for fleas to leave humans after feeding and return to the reservoir species. Those processes, necessary for flea reproduction, would have required time, which was not congruent with the fast spread of the plague during the first two pandemics.

WAS IT FLEAS?

We are running out of possible suspects. What ectoparasites are left? Likely culprits must have been able to tolerate the cold temperatures of Europe, they must have been compatible with the fast spread that marked the first two pandemics, and they must not have been hidden away on the other side of the world when the pandemics raged through Europe. Is there such a suspect? Would we have set up this argument and line of reasoning if we did not have a suspect? Just like in the Columbo series, the culprit was right in front of us, hidden in plain sight.

Lice. Lice? Yes, human lice (*Pediculus humanus*).

The intricacies of body lice, their behavior, the diseases they carry, and their role in history, are discussed in Chapter 8. But it seems that they also provide the likely answer to this mystery. Speculation about *Pediculus humanus* as a vector of the plague began only in the middle of the 20th Century. Despite that early speculation, lice continued until recently to be accorded a minor role at best, with the focus instead on fleas because, well, everyone *knew* that fleas were the vectors. Now it seems the importance of the roles for the two insects should be reversed.

What evidence is there for human lice as vectors?

Two lines of reasoning support this theory. First, the exoneration of other possible ectoparasites, which we just did by eliminating fleas as the vector that caused the fast spread. We did not eliminate fleas from the entire story, simply assigned them very specific roles. Columbo might have considered fleas to be aiding and abetting the crime of spreading the bacterial pandemic. Second, a large number of factors about lice align with the fast spread.

Several insect species are capable of vectoring *Yersinia pestis* among mammal hosts. But of all those possible vectors, the only one that could have been present and numerous enough throughout Europe at the times of both the first and second pandemics was human body lice. A number of studies have shown that human lice are capable of plague transmission. Individual lice, being *much* smaller than fleas, may not be as effective a vector as the Oriental rat flea, but maybe just being present could and should count for something.

A few key points may cinch the identification of lice as vectors. Body lice are specific to human hosts and can transmit diseases directly from one infected human to another. Human body lice originated with and fed on humans in the African Paleotropics and moved with humans to Asia and Europe. Because of their intimate relationship with humans, lice are protected from the elements and kept at temperatures near the human body temperature of 37°C. Also, humans wearing clothes would have ensured that the lice were not subjected to adverse environmental conditions. Clothes and the lack of modern sanitary practices set the stage for human lice to live, thrive, and spread bacteria among human family members and others in close contact, which is consistent with the spread of a disease that often infected all members of one household but left others in nearby houses untouched.

Where are we now? At this point it seems as if getting the plague from the deserts of Central Asia to Europe was like a relay race, a series of vectors and a series of hosts, one after another, passing the baton. Or maybe a better analogy would be the 1987 comedy film *Planes, Trains and Automobiles*, given the different modes of transport of vectors and hosts and the speed at which they traveled.

The exact sequence of insect vectors, mammal hosts, and human infections and fatalities still needs resolution. At times, the spread of the disease was slow, such as during transit along the Silk Roads or on ships crossing the Black Sea or Mediterranean Sea, which is consistent with fleas and rats. At other times the spread was fast, moving at the speed of humans, and not reliant on insect vectors alternating between mammal hosts for reproduction, which is more consistent with lice and human-to-human spread. We can add to the story the narrow windows of time when fleas can transmit the bacteria, as well as the time between when the concentration of bacteria in human blood is great enough so that the bacteria can be picked up by a feeding insect but not so great as to kill its human host. This complex scenario highlights just how many things need to go right for an epidemic to occur. Or, from the perspective of human health, how much must go wrong.

In the previous scenario, we are not cutting Oriental rat fleas out of the picture. We are, however, suggesting their role was very different from the conventional explanations for the first two pandemics. Fleas started the plague's slow movement. Lice almost certainly caused its rapid spread in European and Asian cities, where people lived and died in close proximity. Gerbils were the reservoir for the bacteria, but rats likely carried the fleas, with black rats and brown rats having different roles—black rats on ships or into ports, brown rats in the ports, neither one transporting the disease out of the ports. Humans transmitted the bacteria through close contact, by lice, spreading bacteria and lice at the speed of humans. No need for other reservoir species. Humans were both victims and reservoirs. That simple elimination of the need for an external reservoir is what allowed the clock-speed of human movement to match the rapid rate of the plague's spread seen in both the first two pandemics.

THE STORY CONTINUES

There are still details in need of further clarification to complete this explanation of lice as vectors. Compiling a complete story requires a mixture of genetics, archaeology, epidemiology, tying together loose ends, and deriving conclusions. We still have one more Columbo-esque interjection of "Just one more thing." The question remains how *Yersinia pestis* can survive the high, harsh deserts of Asia in rodent hosts without decimating the rodent populations, which would ensure the demise of the bacteria as well.

Somewhat remarkably, burrowing rodents can pick up *Yersinia pestis* from the soil as they burrow. This is rendered remarkable due to an interesting fact—*Yersinia pestis* cannot survive in the soil. OK. If these bacteria cannot survive in the soil, how are they . . . well . . . surviving in the soil? The soil environment in the high Asian deserts is not really conducive to bacterial survival. *Yersinia pestis* does not survive well when exposed to ultraviolet radiation from the sun. It can survive in rodent burrows, at least in deeper burrows, that can be 3 meters deep. But how do the bacteria survive there? A 2017 study offers insight.[4] The authors found that soil-dwelling amoebae (*Acanthamoeba castellanii* in the study, although this story certainly extrapolates to others) played a key role in enhancing the survival of *Y. pestis*. Amoebae (singular: amoeba), are pretty well ubiquitous in distribution, including in soils. Amoebae consume bacteria they encounter in the soil, by a process called phagocytosis. Basically, amoebae envelop their prey, creating membranous sacs within their bodies—called vacuoles—where prey are killed and digested.

Here's where it gets hyper-interesting. *Yersinia pestis* interferes with the phagocytosis process—the bacteria are engulfed by the amoebae, but not digested. *Yersinia pestis* is able to subvert the digestion process and survive within an amoeba. Instead of being digested within the vacuoles, the bacteria prevent digestion and simply reside in vacuoles unharmed. Vacuoles are seemingly empty spaces in the amoebae, surrounded by a membrane. When the bacteria are inside a vacuole, they are protected from desiccation. As with other parasites throughout nature, it seems that the bacteria take over the process and control the amoeba for their own benefit. And where does one find the amoebae? One place is in rodent burrows. In fact, in the soil in burrows below dead gerbils, where decomposition releases body fluids, moistening the soil and carrying *Y. pestis* with it. Then how do the bacteria get back into other gerbils? As other gerbils root around in the soil, the amoebae containing *Y. pestis* can be inhaled, and the cycle continues.

The bacterial pattern could be viewed as a kind of short-distance migration. The bacteria travel from a dead host into the soil, into an amoeba, and ultimately into another host. Within the amoeba, bacteria replicate, so the individual bacterium that was released into the soil from one dead gerbil is not the same individual bacterium that gets passed on to the next gerbil. This is called multi-generational migration, and is known for a variety of insects. Perhaps the most familiar example is that of monarch butterflies (*Danaus plexippus*), which migrate between the species' summer home in the northern United States and Canada and its winter home in the mountains of central Mexico. Monarchs that fly north in the spring are not the same ones that fly south three to four months later in autumn—multiple generations are involved. The multi-generational migration is a good analogy for *Yersinia pestis*, even though occurring on a smaller scale. Well, it is a smaller scale to us—the migration is still a pretty good distance for bacteria.

This addition to the plague story is both fascinating and frightening. The amoeba-bacteria interaction points out how little we know about interactions on smaller scales, and these scales actually do matter. Processes that occur out of our view often escape our notice and so any recognition of their importance. These interactions also offer additional insight into the complete plague story. Consider the value of scientific inquiry. The complexity of the entire bubonic plague system nearly defies belief. Who would have thought before it began that a study of soil amoebae in Central Asian rodent burrows could have implications for a disease that has killed millions of people for millennia? That is the whole point of basic inquiry—there may be no known or anticipated benefit or outcome, but serendipity happens. All this suggests the conventional model of plague zoonosis needs to be redrawn, to include lice and amoebae. A simple two-dimensional model may need additional dimensions. Also, the complexity of the system ought to alert healthcare officials that emerging diseases—whether new ones or the possible return of previously vanquished diseases—will require vigilance and should not be dismissed as another apparently solved disease problem.

Another aspect of basic science needs to be mentioned. For more than 100 years, the story of how plague spread was built upon the assumption that *Xenopsylla cheopis* is just one species of flea.

What if that is not true? What if the fleas described from different areas and habitats are different species? The taxonomy of this genus—of most fleas, in fact—remains remarkably poorly investigated. We consider it unlikely that the Oriental rat flea is one species distributed from sub-Saharan Africa to the steppe of Russia and the mountains of southeast Asia. It could be, but we suggest this assumption warrants revisiting. This case may be an example of cryptic species—similar in appearance but representing different gene pools with different host ranges.[5] It is difficult to believe that the species of insect implicated in the deaths of hundreds of millions of humans has not been subjected to the rigors of population genetics studies and robust, modern phylogenetic analyses, which would answer the question whether the rat flea is one or many species. The studies and reports of plague were not accompanied by voucher specimens, so it will be difficult now to determine which organism was actually the one involved.

OTHER IMPLICATIONS

The initial center of the Justinianic Plague was Constantinople, but the effects were felt far and wide. Analyses made of the consequences of the first pandemic have yielded results that are all across the board. At one extreme, the pandemic is claimed to have taken a total of 100 million lives, including killing a third of all those then living in Constantinople. Fatality rates of 5,000–10,000 persons per day were claimed for the capital. Justinian's plague has been blamed for the fall of the Roman Empire and food shortages that caused famine for eight years, as well as being associated with great demographic change as people moved around, either to occupy land left vacant following the death of plague victims or even to migrate to areas that seemed safer for survivors.

At the other extreme, different interpretations suggest that what little true evidence can be amassed is both weak and ambiguous. The extant evidence, some would claim, does not support the catastrophic, or maximalist, conclusions: the plague did not cause the demise of the Roman Empire and had far fewer long-term effects on the society of the day. Those interpretations use writings from the time of the pandemic—focusing especially on what was *not* included in them. Mathematical models have been employed to try to figure out which is correct. The truth? Likely somewhere in between.

For the second pandemic, including the years 1346–1353, which were later labeled the Black Death, the estimate of 25 million deaths in Europe is likely much more substantiated. Record keeping, although spotty in some cases and even completely absent in other cases, was still better overall than the record keeping during the first pandemic. The rapid movement of bacteria, reaching all of Europe and the British Isles within just five years, is still, 750 years later, a standard for the spread of a disease. Those details are not questioned. With the increased awareness and gradual acceptance of lice as vectors of the disease, fewer questions remain.

Fatalities were distributed throughout the strata of society. But deaths were not the only consequences of the pandemic. The landed gentry were affected economically as renters died and rent income dropped: less income meant less power. The deaths of agrarian renters also meant crops went unplanted or unharvested, and food shortages piled atop the woes directly attributable to the plague. Revolts ensued as workers demanded more.

The lack of awareness of the cause of the plague and the need to pin the blame on something or someone led to extreme violence directed toward blocs of society, once again targeting Jews. With no understanding of microorganisms or how diseases were spread, seemingly backward ideas were posited as explanations, from astrological planetary alignments to soiled clothes to simply looking a stricken patient in the eye, from "bad air" to arousing the wrath of a vengeful deity. Behaviors adopted by people to avoid or ward off the "pestilence" likewise catch our modern attention. Some people walked around sniffing perfume to avoid the ever-present smell of the dead and dying. Doctors wore oversized masks with nose-holes filled with perfumes or strong-smelling herbs to avoid the smell of death. Some doctors even refused to care for patients, lest they become victims themselves. Family members abandoned their ill siblings, parents, or even children to avoid their own certain death. The fabric of medieval society, based on social structure and the semblance of civility, unraveled.

Despite such examples of poor behavior brought on by the plague, many positive societal changes occurred as a direct result of the pandemic. The loss of agrarian renters gave power to those workers who survived. They were in a stronger bargaining position with gentry and, with increased income from their labor, a middle class of society emerged, which could acquire land and reject the sentence of servitude. Deaths of aristocracy provided openings for the populace to move into positions of decision-making and power. The Catholic Church in Europe lost its absolute authority due to its inability to explain or mitigate the devastation from the disease. Similarly, questions were raised in the Islamic world, as the Prophet Muhammad had promised that the plague would never enter the city of Medina. Although the prophecy of Muhammad and ability of the church to contain the plague both failed, it appears any loss of power to either religion was short-lived.

One other benefit occurred as a result of the plague—the beginnings of the Renaissance. Survivors in the patriciate used their wealth to support artists and writers, extending their patronage to architects, whose medieval cathedrals, mosques, and public buildings were themselves works of art. Having faced and survived the Black Death, the elite had changed their attitudes about life and death, and the art, music, literature, and architecture that define the Renaissance persist today.

Insects—fleas and lice—changed human history in countless ways through their roles in initiating and spreading bacteria that caused the plague pandemics. The impacts were felt throughout major religions and across vast areas of the globe—all of Europe, western Asia, China, India, and Africa. The pestilence that brought death and devastation for millennia also brought sweeping societal changes that benefited humankind. The insects may have been tiny, but their enduring impacts were and are immense. However, realize that fleas and lice did not kill anyone—it was the bacteria.

Oh, just one more thing . . . .

We, too, learned in entomology classes and previous schooling the traditional story of the plague, and we accepted it. Until recent years, the story went unchallenged. The recent findings about interactions between the amoeba and plague bacteria, genomic analyses applied to *Y. pestis*, the analyses of ancient DNA, and fingering human lice for their role, all show us that this story would have been very different if we had written it ten years ago. The story would have been a simple rat-rat-flea-*Yersinia pestis* model, all three pandemics would have been interpreted in relation to that standard model, and all divergent observations, excavations, and epidemiological studies would have been hammered and twisted to ensure the results fit that standard model.

That's ten years ago. What about ten years in the future? With the ability to apply contemporary technology—and even technology we cannot yet dream of—toward remaining questions, the story we would tell ten years from now would be very different, too. That future story will be built upon a combination of old-fashioned running down of clues and making inferences as well as high-tech forensics. However, that future story will still require scientific investigators analogous to detectives wearing rumpled raincoats asking just one more question.

NOTES

1. Finding ancient DNA in the teeth of victims is utterly fascinating. Two references are most relevant: M. Spyrou et al., "Phylogeography of the Second Plague Pandemic Revealed through Analysis of Historical *Yersinia pestis* Genomes," *Nature Communications* 10 (2019): 4470; M. Spyrou et al., "Analysis of 3800-year-old *Yersinia pestis* Genomes Suggests Bronze Age Origin for Bubonic Plague," *Nature Communications* 9 (2018): 2234.

2. The bacterium also acquired two more plasmids (bacterial DNA is organized into many independent segments of DNA, called plasmids). Explaining plasmid function and its connection to plague research is beyond the scope of this book, but we will just state that those intricate details make an already-complicated story even more interesting. For readings about the molecular biology, genomes, mutations, plasmids, and ancient burial sites, see M. Keller et al., "Ancient *Yersinia pestis* Genomes from across Western Europe Reveal Early Diversification during the First Pandemic (541–750)," *Proceedings of the National Academy of Sciences USA* 116 (2019); C. Demeure et al., "*Yersinia pestis* and Plague: An Updated View on Evolution, Virulence, Determinants, Immune Subversion, Vaccination, and Diagnostics." *Genes & Immunity* 20 (2019):; K. R. Dean et al., "Human Ectoparasites and the Spread of Plague in Europe during the Second Pandemic," *Proceedings of the National Academy of Sciences USA* 115 (6, 2018); A. Namouchi et al., "Integrative Approach Using *Yersina pestis* Genomes to Revisit the Historical Landscape of Plague during the Medieval Period," *Proceedings of the National Academy of Sciences USA* 115 (2018).

3. The study of rats was reported in Hufthammer, A. K., and L. Walloe. 2013. Rats cannot have been intermediate hosts for *Yersinia pestis* during medieval plague epidemics in Northern Europe. Journal of Archaeological Science 40: 1752–1759.

4. For the utterly amazing story of amoebae and *Yersinia pestis* that helps explain the bacterium's persistence in the hostile, high-elevation deserts of Central Asia, see J. A. Benavides-Montaño and V. Vadyvaloo, "*Yersinia pestis* Resists Predation by *Acanthamoeba castellanii* and Exhibits Prolonged Intracellular Survival." *Applied and Environmental Microbiology* 83 (2017).

5. The gene pool consists of the population whose individuals have the potential (sometimes separated by distance) to reproduce and spread their capabilities and characteristics, or adaptations.

CHAPTER 7

⌀

The Plague, One More Time

You, someone you love, or someone you know will be affected by an arthropod-borne disease in your lifetime. Rather bold of us to state that, it would seem. But despite all medical efforts to combat them, arthropod-borne diseases still prevail and inspire fear in the 21st Century. Many of them are zoonotic arboviruses (shorthand for *ar*-thropod *bo*-rne virus), which are vectored by arthropods from their natural hosts to humans (e.g., Japanese encephalitis, Rift Valley fever, West Nile fever, yellow fever). Dengue is an arbovirus, but its primary host is humans, not other animals. Not every arbovirus is vectored by insects (e.g., scrub typhus is transmitted by chiggers and Lyme disease by ticks). Some of these viruses cause emerging diseases warranting only surveillance, some have brief, regional flare-ups, others are here to stay. Again, you, someone you love, or someone you know will be affected by an arthropod-borne disease in your lifetime.

Often forgotten and not as eye-catching are age-old diseases such as typhus, malaria, or the plague, all of which are transmitted by insects. Typhus, caused by bacteria (*Rickettsia* and related genera), gets little attention (but see Chapter 9). Malaria, caused by a protozoan (*Plasmodium*), seldom makes history books or the news—perhaps because of its chronic nature, and most reporting focuses on acute, not chronic, events. Malaria, year in and year out, is the greatest killer among all insect-borne diseases, causing deaths of roughly 400,000 people per year, 90% of whom live in African countries. Entire books can be (and have been) written about malaria.

And plague? Plague gets attention when an outbreak occurs because the bacteria—*Yersinia pestis*—can kill large numbers of people, as seen in the tens of millions killed in the three major pandemics. But plague isn't restricted to pandemics and the most recent of the pandemics wasn't the end of the plague. Like a serial killer who disappears between crimes only to surface again and again, the plague didn't go away. The outbreaks linked to the third pandemic continued until the middle of the 20th Century, nearly 100 years after its first emergence in southwest China. Even while

The Silken Thread. Robert N. Wiedenmann and J. Ray Fisher, Oxford University Press. © Oxford University Press 2021. DOI: 10.1093/oso/9780197555583.003.0007

the casualties from the pandemic diminished, the plague and its hosts remained active.

Plague in its natural state is a disease of wildlife that resides primarily in rodents, in what is known as the disease's enzootic cycle. However, occasionally humans are infected when plague spills over from its wildlife reservoir to humans, the zoonotic cycle. Many cases of plague are isolated and do not develop into epidemics, such as recent outbreaks in Mongolia from bobak marmot or rabbit hosts, or in the United States from prairie dogs and squirrels. Western North America is home to a variety of small mammals that are capable vectors of the plague: prairie dogs, marmots, pikas, and endangered ferrets, and every year, cases are reported. Natural outbreaks of plague do occur regularly, even now, in the 21st Century. The World Health Organization (WHO) reported 3,248 cases of plague from 2010 to 2015, with nearly 600 deaths. Most modern cases of plague occur in "hot spots," particularly in poor countries in Africa, India, and South America, because the plague in humans is primarily a disease associated with poverty and poor health infrastructure. Madagascar records plague cases almost every year; in just over three months in 2017, the WHO reported 2,348 cases and 202 deaths there. In Madagascar and other impoverished countries with sporadic outbreaks, efforts are made to prevent the disease from developing into an epidemic.

ATTEMPTED ERADICATION

In an unprecedented historical effort to fight the plague, leaders in the Soviet Union attempted to eradicate it over a vast area.[1] The Northern Caucasus and Central Asian regions are home to numerous rodents that serve as reservoirs for *Yersinia pestis*. Soviet leaders created a plan in the 1920s, later known as the Anti-Plague Institute eradication program, to eliminate rats and other rodents that were plague reservoirs and, at the same time, eliminate the multiple species of fleas that were vectors of the disease (Figure 7.1). The target rodents were primarily great gerbils (*Rhombomys opimus*), several jirds (*Meriones*), and jerboas, which are host to a number of *Xenopsylla*, including Oriental rat fleas (*X. cheopis*).

The Anti-Plague Institute program sought to liquidate the rodents and their fleas, with the major effort beginning in the 1930s, employing tens of thousands of locals in a dual effort—to eliminate rodents, fleas, and plague, but also to provide employment in an impoverished area of the Soviet Union. The workers, including women and young boys, were given the job of putting grain mixed with poisons into rodent burrows. The poisons included chemicals such as chloropicrin and zinc phosphide. Chloropicrin was one of the chemical weapons used during World War I and is highly toxic if inhaled or ingested. Zinc phosphide is a chemical used as a rodenticide, as once the chemical is ingested, stomach acids turn the phosphide into phosphine, which is highly toxic, causing cell death by preventing energy production. Those rat poisons were put into individual rodent burrows by thousands of women and children. By hand.

Figure 7.1 Workers in the southern Soviet Union putting poison in rodent burrows. This photograph was taken sometime between 1930–1940.
Source: Courtesy James Martin Center for Nonproliferation Studies/Middlebury Institute of International Studies at Monterey

The chemical mixtures that were applied in the Soviet Union also included insecticides that targeted fleas living in the burrows. Prior to World War II, the insecticides were inorganic poisons, such as arsenates. After World War II, military aircraft and trucks were equipped to spray chemicals and were used to treat large areas, due to the availability of a new insecticide, dichlorodiphenyltrichloroethane, aka DDT. Part of the larger plan also included burning the grasses to eliminate food sources for the rodents and plowing millions of hectares of land to create new zones of agricultural production.

The official Soviet report from the Anti-Plague Institute stated that plague was eradicated from that vast area. In a 1960 report to the WHO, the Soviets claimed there had been no cases of human plague in the previous 30 years. It is certainly true that no cases had been *reported*. More likely, the official mandate was that there was to be no plague, therefore no cases were officially reported. An unwitting report of plague likely landed its writer in the Gulag. The true story of plague incidence was somewhat different. Although there was less incidence of the disease, outbreaks of plague still occurred in the region every few years—just not officially acknowledged.

The economic costs of the program must have been huge. The human costs had to be greater—exposing thousands of locals, especially the young, to toxic chemicals. Reports of human fatalities from the poisoning program were as common as reports of the plague—in other words, not reported officially. Ecological costs were great in the short term, due to non-target effects, interrupting the food web, and ignoring conservation and environmental priorities. However, rodents, fleas, and plague bacteria returned after only a few years. The attempts at eradication were too ambitious, as the area to be covered was too great. Efforts to prevent infections and epidemics failed—fortunately, in this case, because success would have been devastating to entire ecosystems.

INTRODUCTION TO BIOWEAPONS

There also are those who seek not to prevent or halt epidemics, but to start them. Diseases that can inflict harm or panic, whether in military troops or the general population, can be harnessed and used as bioweapons.[2] Weaponized diseases and their insect vectors are not new or restricted to the 21st Century. Catapults have been used to launch bees at enemies (Chapter 13). As stated in Chapter 5, the early use of catapults by Mongols to hurl diseased bodies into Caffa—if it truly did occur—was not likely done with the conscious thought to initiate an epidemic. Germ theory would not emerge for another 500 years. At most, catapulting corpses alleviated the need to bury the backlog of bodies and passed the problem on to the city's residents. Perhaps it is the gruesome nature of the act of catapulting the dead (and dying) that keeps the story going. Or it may be an inordinate fascination with catapults. Regardless of the veracity of the tale, the later analysis of Caffa and catapults tells of the potential to use diseases as weapons—and to use insects as the transport mechanism. However, the plague was not pursued as a biological weapon for more than 500 years.

Insect vectored diseases were not used in World War I. Why not? Plague and typhus and their insect vectors were known. What alternative weapons did the sparring sides have instead? Gases. Chemicals, such as chlorine, phosgene, and sulfur mustard, were used as weapons. Those gases caused more than 100,000 fatalities and about 1 million non-fatal battle casualties. Descriptions of the troops subjected to poison gases in Western Europe affirm the horror of those new weapons.[3]

The numbers of military casualties and fatalities from chemical weapons pale by comparison with the totals of 8.5 million fatalities and 20 million casualties incurred in battle. Even so, after World War I, the world recognized the horrors of chemical warfare, and so the 1925 Geneva Protocol was developed as an international treaty to prohibit the use of chemical weapons that caused poisoning or asphyxiation. Although disease-causing pathogens were not used as biological weapons at the time of the 1925 treaty, the protocol included bacteriological methods in its list of banned agents of war, recognizing the potential for their use. Most of the world's major countries ratified the treaty, though not all. There were two primary exceptions.

Disease-causing pathogens were not used in World War I, but both Axis and Allied powers tested and deployed pathogens before and during World War II. Great Britain experimented with anthrax, and Germany exposed concentration-camp inmates to typhus, carried by lice (discussed in Chapter 8). The United States investigated a number of pathogens that could be vectored by mosquitos, lice, and fleas. Although no evidence exists that it deployed insects to vector pathogens, a large wartime expenditure was used to establish research and production facilities to produce insects and pathogens that could be used as weapons. From 1938 to 1945, the major players in biological warfare—specifically entomological warfare, because of the use of insect vectors—were three of our Silk Roads countries: the Soviet Union, Japan, and China, as investigators, aggressors, and primary victims, respectively.

Soviet efforts may not be fully known, but it has been reported that they used both prisoners of war and political prisoners as subjects for experiments with bubonic plague. In a remote area of Mongolia, prisoners were kept in tents containing plague-infected rats and rat fleas. As the rats died, the fleas moved to the detained humans for blood meals. However, at least one of the captives escaped, from which a plague outbreak started. Fearing world exposure, the Soviets bombed and burned Mongol villages in the vicinity to prevent an epidemic as well as to prevent escape of information about their experiments. After the war, Soviet bioweapon efforts increased greatly, with dozens of laboratories scattered across the Union. For the record, the Soviet Union was a signatory to the 1925 Geneva Protocol. The 1925 Geneva Protocol didn't prevent testing of biological agents, it prohibited deployment of them as offensive weapons.

JAPANESE BIOWEAPONS: DISEASES AND INSECTS

The Japanese program to use insects to spread disease made all the other countries' efforts look amateur. Concerns that not every country would sign the 1925 Geneva Protocol (Japan did not ratify the treaty) led Japan's Imperial military leaders to believe that biological weapons would be deployed by military powers against Japan, and so they acted first. The genesis of Japan's bioweapons program was their 1931 invasion of Manchuria, in far northeastern China, and establishment of the puppet state of Manchukuo. Japan had already been granted, in 1906, control of the mainland-based South Manchuria Railway, which was part of the Chinese Eastern Railway network, linking the eastern terminus of the Silk Roads with northeastern China, Russia, and Korea and extending the Silk Roads to Japan, much as the link to Japan extended silk production there, in about 300 CE.

To acquire the global power it sought, Japan required raw materials to forge military machinery, many of which were to be found in northeastern China. Occupation of Manchukuo was the first part of Japan's plan for domination of Asia. The occupation led to tensions and border skirmishes that mushroomed into the Second Sino-Japanese War, beginning in July 1937. Some historians consider 1937 and the Sino-Japanese War to be the actual beginning of World War II. The Second Sino-Japanese War resulted in deaths of more than 20 million Chinese civilians and some of the most brutal treatment and atrocities ever inflicted on civilians.

Prior to the war with China, the Japanese military had begun a biological weapons program, initially based in Tokyo and led by General Shiro Ishii, a surgeon and bacteriologist. As the program grew, its danger to the Japanese homeland was realized, so the program was moved to the Asian mainland, first to the city of Beyinhe and then to nearby Harbin, in the occupied state of Manchukuo. Known as Unit 731, the program's official name was the "Epidemic Prevention and Water Supply Unit of the Kwantung Army"—ironically named, given their mission was to *cause* epidemics, not prevent them.[4] With the creation and occupation of the state of Manchukuo, General Ishii's program could be safely located far away from Japan, and a network of satellite units was developed.

Further growth of Unit 731's program required a more secure site, so a move was made to a large, purpose-built, permanent facility at Pingfan, near Harbin (Figure 7.2). The site was 6 square kilometers in area and contained some 150 buildings. At its peak, Ishii's greatly expanded program employed as many as 10,000 workers, half of whom worked at Pingfan. By 1938, the composition of the staff had shifted from military to physicians and prominent medical researchers who previously had worked at Japanese universities. The vast scope of human research conducted by Unit 731 personnel reflected the disdain held by Japan for the Chinese and the depravity with which the Japanese acted against Chinese civilians.

Unit 731 personnel at Pingfan worked around the clock to produce various bacteria in huge containers, as well as to grow insect vectors in an insectary. The bacteria produced included *Vibrio cholerae* (which causes cholera), *Salmonella enterica* (the serovars, which cause typhoid fever and paratyphoid), and *Bacillus anthracis* (which causes anthrax). But, even with other bacteria in production, the major focus of the Pingfan facility was producing large quantities of *Yersinia pestis*. What do we mean by large quantities? Roughly 300 *kilograms* of viable plague bacteria were produced

Figure 7.2 Building on the site of the bioweapon facility of Unit 731, near Harbin, China.
Source: 松岡明芳/CC BY-SA 3.0/Wikimedia Commons

monthly. Monthly. Sounds like a lot, but how much is it? Let's do the math. Estimates of the mass of an individual bacterium range from 10^{-10} to 10^{-15} grams. That's a pretty broad range, but all of these numbers are less than miniscule. It would take a lot of bacteria to build a skyscraper. For our example, if we take the midpoint, 10^{-12} grams, 300 kilograms would equal approximately 3×10^{17} bacterial cells or, in words, approximately 300 quadrillion individual bacteria. Produced every month.

General Ishii had studied the pandemics, and so he believed that the best route for vectoring plague was insects. A large granary was built at Pingfan to feed and house rats, needed as hosts for the flea vectors. At the program's peak, an estimated 3 million rats were used to produce 100 million fleas *every few days*. So, the Japanese had the players in place: plague bacteria, fleas, and rats. All three were there in huge numbers: 300 quadrillion bacteria per month, 3 million rats, 100 million fleas every few days. What they lacked was a means to deploy their weapon system (and we are grateful they did not know to use lice!).

UNIT 731

In October 1940, Japanese bombers attacked the Chinese city of Ningbo, but not with a conventional payload. Flying at an altitude of only 200 meters, bombers dropped plague-infected fleas, along with grain and pieces of cotton. The fleas had fed on infected rats and were deployed as vectors, under the assumption that infected fleas could harbor the bacteria safely (safe for the Unit 731 personnel, anyway) and could deliver bacteria to hosts more effectively than could be done by simply dropping bacteria. An epidemic began within a week, but its symptoms were recognized as plague by locals, who then took steps to mitigate any further spread of the disease. After six weeks, the epidemic had abated after causing about 100 deaths—hardly the results expected by General Ishii. The mission was considered a failure.

The Unit 731 command realized that deliberately spreading disease was not as easy—or as easily controlled—as had been thought. Other attempts resulted in a few infections, but no spread and no epidemic. In one offensive in 1941, Unit 731 aircraft delivered bacteria with the intention of causing an epidemic of cholera among Chinese troops, dropping the bacteria onto a battle zone. Unfortunately for the Japanese, the advance of their own Imperial troops in the battle took them through the drop zone. The subsequent 10,000 cholera infections and 1,700 deaths would have been useful to their effort except for one thing—all victims were Japanese troops.

After the cholera debacle and with waning support among military leaders for biological weapons, it was clear that a better means of delivery was needed. General Ishii suggested a "bomb" containing the bacteria with or without insects. Standard explosive bombs would not do—the explosion and heat would kill the enclosed bacteria and insects. General Ishii was not deterred. Not only was he a physician and bacteriologist, he apparently was also somewhat of an inventor, having previously invented a portable water purifying system. General Ishii developed a ceramic "bomb" that either could be detonated at low altitude with a small charge, thus spreading the insects and bacteria, or could be deployed unarmed and would break apart on impact, releasing hungry fleas from its ceramic shell, thereby dispersing them to bite any nearby humans.

The "Ishii bacteria bomb" was used as a delivery device in hundreds of bombings in both manners for the remaining years of the war. In one incident, the Japanese used a two-part bombing routine to initiate an epidemic of cholera in the Yunnan Province of southwestern China. For the first part of the plan, traditional explosive bombs were dropped along with ceramic bombs with two chambers: one held a yellow slurry of the bacteria that cause cholera; the other chamber held houseflies—the houseflies had been grown on waste from the rats used for plague experiments. The explosive bombs killed thousands of civilians. The "unexploded" bombs broke open on impact, thus exposing the flies to the bacteria. The flies distributed themselves among nearby people, thus initiating infections. A few days later, a conventional bombing run caused people still in the city to flee to the surrounding countryside, thus carrying the disease with them and initiating an epidemic that allegedly killed as many as 200,000 Chinese civilians.

Still other methods to enhance infections illustrated the intersection between medicine, insect science, and depravity. Citizens in Chinese cities were starving. To capitalize on their suffering, the Japanese dropped ceramic bombs along with grain containing infected fleas. As hungry people converged on the grain, they were bitten by the fleas and infections ensued. In other cases, pieces of colored paper covered in plague bacteria were dropped, to attract the interest of children. Again, plague infections began, in this case without the need for fleas, at least not to initiate the infections.

DESPERATE MOVES

Plans were made by the Unit 731 leaders to take the bio-warfare elsewhere and use the plague bacteria against Americans, both combat soldiers and civilians. In 1944, a Japanese vessel carrying an assault team was dispatched to Saipan, by that point an important airbase for B-29 bombing missions targeting the Japanese homeland. The plan was to release infected fleas near the US airfield to initiate an epidemic among US air crews and other soldiers. But, while en route to Saipan, the ship carrying the assault force and lethal cargo was found and sunk by a US Navy submarine before the mission could be completed.

In a plot that sounds like a Jules Verne science fiction story, long-distance balloons were envisioned as transoceanic delivery devices. The balloons had been tried in a failed plan to start forest fires in the United States, but the same balloons were also considered for use in a plan to transport plague bacteria across the Pacific Ocean. Those plans never manifested into actions. In a more-conventional plot begun near the end of the war, a Japanese submarine was dispatched carrying a small aircraft that could be launched from the ocean surface. The plan was that the plane would disperse plague bacteria over southern California, to initiate an epidemic and cause panic. Allegedly, the Japanese leadership, fearing US retaliation, intervened at the last minute and the mission was aborted.

By the middle of 1945, the end of the war was looming. When the Soviet Army invaded Manchuria, General Ishii was forced to destroy the Pingfan facility (and the other satellite facilities), leveling and burning buildings and, with them, any incriminating records. As the granary was destroyed, large numbers of rats, carrying infected rat fleas, escaped. Almost immediately, a plague epidemic occurred in the surrounding area as the rats found shelter and the fleas found human hosts.

The numbers of fatalities inflicted by using insects to initiate and spread plague bacteria are unknown. Estimates, difficult to substantiate, range upward of a few hundred thousand Chinese citizens killed, though those numbers contain the fatalities caused by other diseases, such as cholera (spread by flies released with the bacteria), typhoid, and dysentery. Because of the lack of surviving records and sketchy details provided by Unit 731 staff after the war, the true numbers will never be known. But the numbers matter less than the fact that such atrocities were perpetuated, and unfortunately insects were used as a means to cause such horrible crimes. The actual number that matters is *one*. One victim, one fatality, one ceramic bomb was too many, and one fatality was sufficient to demonstrate a depravity that should never have been seen.

After the War, most Unit 731 personnel returned to Japan. Those who did not were arrested by the Soviets and tried as war criminals. However, the Unit 731 leaders and other personnel who returned to Japan escaped accountability and justice. General Douglas MacArthur, on advice of his staff and US civilians who interrogated the unit's captured personnel, promised not to prosecute them if they provided information. The knowledge provided by the unit's scientists was seen as valuable to the only other country that failed to ratify the 1925 Geneva Protocol. The United States, with its growing biological warfare facility at Fort Detrick, Maryland, and with its production of potential bioweapons at converted arsenals scattered around the country, wanted the information, and so those perpetrating these war crimes were not held accountable.

NOTES

1. An excellent and richly developed report of the Soviet's anti-plague efforts is found in: S. D. Jones et al., "Living with Plague: Lessons from the Soviet Union's Antiplague System." *Proceedings of the National Academy of Sciences USA* 116 (2019). Much of the material about anti-plague efforts presented here comes from this article. Additional information appears in: Sarah Zhang, "The Soviets Tried So, So Hard to Eliminate the Plague," *The Atlantic* (2019) https://www.theatlantic.com/science/archive/2019/05/when-soviets-tried-to-eradicate-the-plague/589570/. A wealth of information is available in various reports by the James Martin Center for Nonproliferation Studies at Middlebury Institute for International Studies in Monterey, California, https://www.middlebury.edu/institute/academics/centers-initiatives/nonproliferation-studies.

2. The definitive source of information about use of insects as weapons is: J. A. Lockwood, *Six-Legged Soldiers: Using Insects as Weapons of War* (New York: Oxford University Press, 2010). Lockwood's encyclopedic coverage of weaponized insects begins with ancient cultures and continues through the war in Vietnam, and his book

goes far beyond what we can present in the current book. The book reads like a novel, but better—it is informative and awesome.

3. To gain a better appreciation for the devastating damage caused by chemical weapons, we recommend: Wade Davis, *Into the Silence: The Great War, Mallory, and the Conquest of Everest* (New York: Vintage Books, 2012). His storytelling includes the effects of World War I on Great Britain and the drive to retain the luster of British Empire after the loss of most of a generation to the war. The tale alternates between treks in the lofty peaks of the Himalayas and the attempted conquest of Mt. Everest by George Mallory, Sandy Irvine, and their party, and brutal accounts of battles on the Western Front. Davis heard stories of the war from his grandfather, who was a medical officer in the field. Davis's account of the gas warfare reinforces why the Great War should have been the "war to end all wars."

4. Numerous books and other publications detail the extent of the atrocities committed by Unit 731 personnel. We are not including them in this book, as they do not contribute to the entomological story presented. Readers wishing to pursue the subject will find far more references and material than is necessary to incite revulsion.

SECTION 3

Lice in War and Peace

—430 BCE Plague of Athens
—1546 CE Typhus first described
—1780 CE John Howard starts prison reform, Howard Association formed
—1836 CE William Wood Gerhard distinguishes typhus from typhoid fever
—1845 CE Potato blight destroys potato crop in Ireland
—1846 CE Emigration of Irish to avoid Great Hunger, typhus kills 20,000
—1848 CE Rudolf Virchow establishes idea that "medicine is a social science"
—1903 CE Nicolle determines body louse as carrier of typhus
—1909 CE Ricketts discovers causative bacterial agent of typhus and dies from it
—1914 CE Typhus epidemic in World War I kills 150,000 in Serbia
—1915 CE Prowazek studying bacteria, dies. Rocha Lima names *Rickettsia prowazekii*
—1918 CE Russian Great Epidemic begins, kills 5 million in five years
—1941 CE Typhus epidemic averted in the Warsaw ghetto
—1942 CE Battle for Stalingrad begins, 90,000 Axis prisoners captured, typhus kills most
—1943 CE Typhus outbreak averted in post-War Italy

CHAPTER 8

∽

Lice in War and Peace

"Quit being a nitpicker!"

"You nitwit!"

"I feel lousy!"

Phrases like these are commonplace in casual speech, and they have a common origin: small insects called parasitic lice. "Nits" are the eggs of lice. They are small, oval-shaped, about the size of a pinhead, and are typically not desirable in your hair. Perhaps you have seen images of gorillas or chimpanzees sitting together, in a grooming "salon," fingers carefully threading through the hair of another, picking off (and often eating) the tiny nits they find. Analogously, to "nitpick" is to criticize or pick out minutiae. Most phrases like this have similar origins. "Lousy" is a little more involved. "I feel lousy" refers to feeling ill—a direct nod to the knowledge that lice spread disease (although it has expanded to also include general badness, such as "He played a lousy game tonight" and "She is a lousy poker player"). Although this may seem like obvious common knowledge today, lice were not always known to transmit disease. But before we can dive into disease transmission, we must first say something about what lice actually are. So, let us first examine with a fine-toothed comb this tale we've "nit" together and hope we don't do a lousy job . . . .

Sorry about that. Moving on . . . .

INTRODUCING THE AMAZING LOUSE

Despite the revulsion most people feel toward lice, they are fascinating insects, with diversity and specializations that are truly amazing. Revulsion stems from those two species that affect humans, but there are well over 11,000 species of lice organized in a single order (Psocodea), spanning a breadth of habits. Roughly half of known

The Silken Thread. Robert N. Wiedenmann and J. Ray Fisher, Oxford University Press. © Oxford University Press 2021.
DOI: 10.1093/oso/9780197555583.003.0008

lice are free-living species often called bark lice or barkflies. These can be common insects found in a variety of familiar habitats like backyard trees, dust piles, and bookshelves. Some are quite weird, including species in which the female has a penis that she uses to penetrate the male to acquire sperm. In fact, this bizarre behavior evolved twice—once in South America (*Neotrogla*) and once in Africa (*Afrotrogla*)! Other species are attractive: one resembles tiny, quick-flying moths sometimes seen on exposed forest rocks (*Stimulopalpus japonicus*); some live together as families in huge webs (*Archipsocus nomas*); and still others (i.e., booklice) are associated with humans and can even be pests (e.g., *Liposcelis bostrychophila* and *Trogium pulsatorium*). Our story involves the other species, the parasitic lice.

Parasitism evolved only once among lice, sometime between 100 and 115 million years ago, at the end of the Early Cretaceous. All of today's 5,000 species of parasitic lice are descendants of that one common ancestor and all are ectoparasites of birds and mammals. The parasitic lice are part of the Psocodea, organized into a group called Phthiraptera (formerly treated as an order). Most species of parasitic lice are "chewing lice," named for mouthparts that allow them to feed on their host's dry skin and debris (some actually feed on blood that pools as they chew). Some chewing lice use mammals as hosts, but most are parasites of birds. "Chewing lice" is an informal designation that basically refers to all the parasitic lice that are not sucking lice. Sucking lice, by contrast, comprise only 500 species that feed exclusively on mammal blood by piercing skin with specialized needle-shaped mouthparts. Sucking lice, unlike chewing lice, are all each other's closest relatives and are organized into a group called Anoplura (formerly treated as its own order). There are many interesting sucking lice, but our story involves species that parasitize humans.

Humans are hosts to two sucking lice: pubic lice (*Phthirus pubis*) and human lice (*Pediculus humanus*)—the latter of which comprises two subspecies: head lice (*Pediculus humanus capitis*) and body lice (*Pediculus humanus humanus*, Figure 8.1). These two subspecies sequester themselves on different parts of their host's body. Our bodies. Head lice are found exclusively on the heads of humans, and body lice feed on the rest of the body. But body lice actually live on clothing worn by humans and move onto humans only long enough to feed. Human head lice are those that get passed among children in day care or elementary schools,[1] and that result in the dreaded notes from a teacher, but they are not disease carriers. It is the human body louse that is the vector of diseases and the one that has impacted human history enormously.

AN UNEXPECTED TANGENT

Studies on the evolution of our three kinds of lice bring into view aspects of our own evolution. To illustrate this, let's briefly go off on a tangent into a story about pubic hair and gorillas, because some stories are just too weird to withhold. Pubic lice (*Pthirus pubis*) share a common ancestor with gorilla lice (*Pthirus gorillae*) and that common ancestor likely also parasitized gorillas, meaning we acquired pubic lice

Figure 8.1 Female body louse, *Pediculus humanus*.
Source: Photo by James Gathany, courtesy of the Centers for Disease Control. Image 9215.

from gorillas in Africa. The likely answer to how that occurred tells us something about our own past as well as our present bodies.

Humans are thought to have acquired pubic lice from gorillas about 3.3 million years ago, around the time that Lucy—a three-year-old *Australopithecus afarensis*—fell into a rushing stream and died. Lucy's fossilized bones are famous and remain among the most complete fossils of our ancestors, offering us the chance to examine our beginnings. Much attention about our past has focused on bipedalism, brain size, jaw size, diet, and family structure. But what about pubic hair? Or body hair in general? After all, her hair did not fossilize. We now know *Australopithecus* lived at higher elevations with lower temperatures. Studies highlighting the importance of body hair to thermoregulation make it obvious that hair loss could not have evolved until much later, when early members of our own genus (*Homo*) began exploring open areas at lower elevations, with a hotter climate, about 2.5 million years ago. Even then, we weren't hairless, having hair on our heads, sexually dimorphic beards, and pubic hair, which marked the age of sexual maturity (i.e., the age when pubic lice could be acquired).

Understanding why we have pubic hair is essential to understanding why we have pubic lice. Our body hair is scant, but for the bit around our genitals. What about other apes? They have hairy bodies with only finely haired genitals. We seem to be inverses of each other. In other words, our pubic hair is not a leftover from when we had hairy bodies; rather, it is a specialized structure that affected our evolution. What is its function? Many ideas have been proposed, but the currently accepted one involves odor. Yep. Pubic hair once functioned as a pheromone duster (and maybe still does!), allowing us to better detect pheromones that conveyed clues about the genetic fitness of sexual partners. And these pheromones were so important to our reproductive advantage that we evolved a structure that allowed us to better smell them. Why does this matter? Because all the while that we were assessing our partners, we were also passing around pubic lice. OK, got it. There's just one problem . . . .

Remember, *Pthirus pubis* diverged from *Pthirus gorillae* roughly 3.3 million years ago. This was almost a million years before early *Homo* even evolved pubic hair. Where did pubic lice live? Could it be that *Australopithecus* had pubic hair? Unlikely. Far more likely is that pubic lice once resided across a broader area of the body than just the genitals, but through time, as that lineage lost body hair, pubic lice were relegated to the pubic region, and that would render pubic lice, not pubic hair, the present-day leftovers of our hairy past. And there you have it—human sex pheromones, gorillas, and philosophy, all stemming from pubic lice. Ok, tangent over, back to body lice . . . .

In contrast to pubic lice, human lice (*Pediculus humanus*) shared a common ancestor with chimpanzee lice (*Pediculus schaeffi*) 6–7 million years ago: rather than being part of a complex story of host switching, these lice simply never left us. They were with us when we looked a lot more like chimps, and they witnessed our standing upright, losing our body hair, adorning ourselves with clothes, and developing complex culture. Human head lice and body lice that are found in Africa both have greater genetic diversity within their species than lice found elsewhere, which further supports the theory that humans originated in Africa. Species found in their ancestral home generally have greater genetic diversity than when they have moved elsewhere. The two subspecies of human lice diverged roughly 100,000 years ago, which corresponds to the time when humans began to wear clothing. Not Armani suits, but animal skins. And as we shrugged on our attire, some human lice—which had been relegated to head hair—retreated into clothing for safety after feeding.

Wearing clothes enabled humans to move out of Africa and into the cooler regions of Eurasia. Early clothes were restricted to animal furs, as the means and knowledge for using plants to form fabrics or clothing would not be acquired for eons. Clothes not only kept humans warm, but also kept infesting body lice near the optimal temperature for humans: 37°C. By wearing clothes to keep warm, humans make life for a louse much easier. No lousy life for a louse.

Head lice are easily spread in such things as hats and coats that are in direct contact with head hair. Body lice, living in or on clothing, are spread by sharing clothes or being in close proximity, so that the lice can be transferred to the clothes of others. Both body and head lice have come to be associated inappropriately with poverty or unsanitary conditions, thus the revulsion. Head lice may have the "yuck-factor," but body lice have their own, different "yuck-factors." Body lice transmit diseases—nasty diseases—several of which have greatly impacted history.

BODY LICE

Body lice are now considered to have been a major reason for the rapid spread of bubonic plague in the two pandemics of the 6th and 14th Centuries, serving as the primary vector for human-to-human transmission (Chapters 4–7). Although the traders along the Silk Roads almost certainly were infested by body lice, those lice likely did not start the plague infestations in Southern Europe and Western Asia, but rather supported their spread once the bacteria reached European and Asian ports. However, human body lice are not notorious for their role in spreading the plague (though that is changing as we write this) but for spreading typhus-causing bacteria (*Rickettsia prowazekii*). Because this disease is often manifested in the wake of natural disasters or wars, resulting in rapid spread, it is referred to as epidemic typhus. Different kinds of typhus are vectored by fleas (murine typhus) or chiggers (scrub typhus). Other diseases transmitted by body lice include trench fever and relapsing fever, each caused by different bacteria.

Unlike other insect vectors of diseases, such as fleas (Chapters 4–7) or mosquitoes (Chapters 9–12), lice transmit disease through their feces (called frass). AKA, bug poop. Bacteria grow in the gut of a louse, and when the insect defecates, typhus bacteria are excreted. The bite of a body louse on a human host itches, and scratching ensues.[2] But usually there is more than one louse present. Infestations by body lice can reach into thousands of lice, and each louse bites its human host several times per day. So, do the math—multiple thousands of bites per day, causing severe itching, and scratching. When the bitten human scratches the itch, bacteria in their feces get rubbed into the bite wound, or into a wound created by severe scratching. Bacteria in dried louse frass are still viable and virulent for days after the adult defecated. *Rickettsia prowazekii* will kill the louse, but bacteria remaining in the insect's gut are capable of causing infection in humans even then.

Typhus is not very forgiving, nor does it act slowly. Once a bite becomes infected, the onset of disease requires 1–2 weeks of incubation in the human host. Symptoms of typhus usually begin with fever, reaching 40°C, followed by a rash that begins in the chest or abdomen, spreading to the arms and legs. Symptoms can advance to severe muscle pain, dangerously low blood pressure, nervousness, insomnia, and delirium, followed by renal failure and death. Without treatment, mortality from typhus is generally 30–40%, but rates can reach 100%.

Typhus is not a new disease, though its early history is difficult to pin down. Ancient reports of an epidemic in Greece in 430 BCE, called the Plague of Athens, have been analyzed, and the reports of the illness are consistent with typhus. Reports of disease in Europe through the first half of the second millennium may have been typhus, though diseases with similar symptoms, such as typhoid fever or plague, may also explain the epidemics. The formal description of typhus was issued in 1546, but even three centuries later physicians still confused typhus and typhoid.

It was during the 1836 typhus epidemic in Philadelphia that this disease was finally distinguished from typhoid by American doctor William Wood Gerhard, who realized through first-hand observations of patients—both pre- and post-mortem— that the two diseases were very different. Although the epidemic patients he examined had high fever, they lacked intestinal problems, which was characteristic of typhoid. Therefore, the epidemic was not caused by typhoid—so, then, what was it? Interestingly, Gerhard noticed that cases occurred in clusters from the same residence, largely from certain neighborhoods—those known for their vice and poverty. Rather than focus on the poverty in the neighborhoods, Gerhard realized that it was the density of residences or their inhabitants, not their impoverished status, that correlated with the disease. He attributed the disease's spread to proximity rather than poverty. His observations and inferences represented a tremendous leap in epidemiology, especially in his examination of all the evidence as it existed—many of his contemporaries looked at evidence only after a conclusion had already been drawn. Despite Gerhard's insight into alternative perspectives and advances in epidemiology, it totally escaped his notice that lice might be playing a significant role.

JAIL FEVER

Because of the confusion among diseases with similar symptoms, attributing to lice the cause of ancient epidemics is no easy feat. Documentation of typhus really began in 18th-Century England, in jails. First, let's differentiate between jails and dungeons. Dungeons were the dark, dank spaces in the depths of castles where prisoners from battles were kept. "Throw him into the dungeon!" is a line we all probably heard when we were children, conjuring a mental image of a prisoner shackled to a wall, fed gruel (whatever gruel was), and whipped with a cat-o-nine tails. The tale was likely more myth than reality, though use of dungeons for prisoners certainly did occur.

Jails were a different thing entirely. Other than debtors' prisons, English jails were little used until the 17th Century. Debtors' prisons were paradoxical because the imprisoned had to pay for their food and "lodging," making no sense because the debtors were imprisoned due to their inability to pay their debts—how, then, were they to pay for room and board? Other than for debts or very minor infractions, the judicial process was pretty simple: a guilty plea or verdict from the court resulted in the death penalty; an innocent verdict resulted in being set free. Penalties for petty crimes included time in the stocks, public whipping, or even being submerged in water on a ducking stool. That was for petty crimes. Other crimes were dealt with differently. Realize that, at that time, something like 240 crimes carried the death penalty—240 of them!

At the time, jails or prisons held the accused until trial or those found guilty and awaiting their sentence. Prisons were not used for punishment until after 1605 and the publication of *Utopia* by Thomas More, in which he proposed imprisonment as an alternative to the death penalty. Imprisonment represented a real alternative for dealing with criminals. It was a great idea, but there were not enough prisons. Enacting More's proposal required building or renting them. Early prisons were privately owned, and state-owned prisons didn't come onto the scene until the late 1770s.

Concern for the welfare of prisoners was likely minimal, given that the primary sentence for a guilty verdict was execution. Consequently, prisons really were like the mental images we have of them: dank places, crude, overcrowded, with little or no sanitation, operated by often-sadistic jailers. Men, women, young boys and girls, murderers awaiting execution, and those accused of minor offenses were all thrown together into poorly maintained facilities. Eventually concern about these conditions led to prison reforms, such as those proposed by John Howard in the late 18th Century and by members of the eponymous Howard Association, but reforms were slow to come. Meanwhile, complementing land-based jails were "floating prisons"— overcrowded ships—anchored in London's Thames River. Or the equally crowded ships were used to transport shackled prisoners to distant colonies, such as Australia or America, where they often were sold as indentured servants. Conditions on the floating prisons were no better than in the land-based prisons, and concern for prisoners was equally lacking.

The irony is that a jail sentence, or just being held for a trial, was in fact a death sentence. Many 17th-Century humans were infested with body lice. When lice-infested prisoners were thrown into crowded conditions, infestations spread to others. Any lice that carried the typhus pathogen spread the disease, which could run rampant through a jail. A colloquial name for typhus was "jail fever," aka "gaol fever." As many as 25% of those in prisons died of jail fever, and allegedly, more prisoners died from jail fever than were executed for all the 240 crimes that called for the death penalty. Lice-infested prisoners hauled into court were known to infest guards, sheriffs, and even judges, through proximity and movement of lice among articles of clothing. At the time, of course, the role of lice in transmitting the fever had not been determined. Thus, a second irony: the magistrate handing down a death sentence may have received the same for himself.

LICE AND THE GREAT HUNGER

Lice figured prominently in the story of the Irish famine and the emigration of the Irish to the Americas. Potatoes were the staple food of Ireland in the 1840s and, for many, almost the sole food. According to author John Keating, in his book *Irish Famine Facts*, the average working man consumed more than 14 pounds (6.4 kilograms) of potatoes each day, adult women more than 11 pounds (5.0 kg).[3] To make those numbers more real, that was 40–60 potatoes consumed *per day*, or more than 2.5 tons of potatoes consumed every year by every working man. Potatoes were grown on nearly any land, in small plots, and in poor soils. Roughly one-third of all Irish potatoes were dedicated to feeding cattle, which were the primary agricultural export to England. Only one variety of potato, the "Irish Lumper," was grown in Ireland. That narrow source of genetics for the potato plant left the country's major crop susceptible to infection, in this case potato blight, a disease caused by water molds, namely *Phytophthora infestans*.

The infection of potato crops began in 1845, killing more than one-third of Irish potato plants. The potato blight worsened in 1846, destroying more than 75% of the potatoes grown, virtually a complete collapse of the crop. The loss of this significant source of food for peasant tenant farmers resulted in what is known as the Great

Hunger. The famine was worsened by government policies that demanded Ireland continue to export its grain crops of wheat, oats, and barley, plus the aforementioned cattle, to England (of which Ireland was then a colony), which had to be done under armed guard. During the Great Hunger, an average of six ships left Ireland with food exported to England for every ship that arrived with aid from England. While Ireland starved, England was indifferent or resentful, with shrill newspaper opinion pieces decrying any thoughts of economic relief. Even in the middle of the accumulating Irish deaths, some English officials considered the famine to be divine intervention to stop the growth of Ireland's population. When England finally relented and offered food aid, the grain shipped to Ireland went uneaten because of a shortage of facilities to mill the grain—and still the exports of food to England continued.

Within five years of the arrival of the blight, more than 1 million Irish had died of starvation and diseases that were worsened by malnutrition. During the period 1846–1851, close to 2 million Irish sought to emigrate, mostly to North America. In the largest movement of people in the 19th Century, poor, starving emigrants tried to escape the famine—almost 25% of the country's population. The emigrants didn't know what they would find across the Atlantic, but they knew that they were leaving behind certain death. Unfortunately, their escape from one tragedy placed them squarely in the middle of another, and that tragedy was caused by lice.

The ships carrying emigrants were crowded with destitute passengers, those who had been referred to as "the unwashed masses." Many were shipped overseas because they could no longer pay for the meager land they lived on. Poor emigrants were already weakened by malnutrition and infested with lice. Although lice were an inevitable part of life for nearly all people in the 19th Century, the "unwashed masses" with unwashed clothes were especially heavily infested. And, unseen, some of the lice carried by the emigrant passengers had their own passenger—bacteria that cause typhus.

In just the first two years of the famine, more than 100,000 Irish left their country: these numbers overwhelmed the capacity of emigration officials. Passengers were given a medical "exam" prior to boarding a trans-Atlantic ship, but the exam was less than cursory. A doctor looked at the tongue of each emigrant as she or he passed by a window behind which the doctor was seated. That was it. Was it any wonder that bacteria and disease boarded with the passengers? Conditions aboard the vessels were bad beyond belief. One ship, the *Elizabeth and Sarah*, was built in 1763 and, for her size, should have carried only 155 passengers. The 83-year-old ship departed in July 1846 with 276 passengers on board (the ship's manifest listed 212). Of the 36 berths on board, 4 were reserved for the crew; the remaining 32 were shared by the 276 passengers. There were no sanitary facilities. None. Emigrants were crowded together in holds with no ventilation, the only light entering through an open hatch.

Many of the ships into which the emigrants crowded had been built to carry timber or to travel between coastal cities and were never meant for transporting passengers, let alone for ocean crossing. Their voyages across the Atlantic should have taken about six weeks—the fastest crossing was 27 days—but frequently lasted 8–9 weeks. One lasted 74 days. The ships' owners planned for only enough food and water for passengers for the shorter sailing time. The Passenger Act of 1842 required that passengers be given 7 pounds (3 kg) of food per week, but the law was unenforceable once the ships were on the sea. Some passengers brought a small amount of food with them, such as salted fish, but most had nothing and relied on the legal provision of weekly bread. On many ships, the food that had been promised for the journey was never distributed, thus worsening the plight of those already malnourished. When bread was provided, it was often moldy and much of it went uneaten. Water was scarce, and passengers were always thirsty. The salted fish were thrown overboard, as they worsened the endless thirst. The *Elizabeth and Sarah* should have carried more than 12,000 gallons of water based on the number of passengers on board. Only 8,700 gallons of water went aboard, and many of the casks leaked.

COFFIN SHIPS

The transport vessels carrying poor Irish across the Atlantic in 1846 and 1847 became known as "coffin ships," because for many passengers leaving Ireland and seeking refuge abroad, the hellish voyages were deadly. When "ship fever" broke out, the sickened were confined to an unventilated hold, in the dark, with no sanitary facilities, no food, and no water. One account estimated 8,000 passengers died en route, but that estimate is likely to be too low. More than 400 ships—later called the "fever fleet"—disgorged their human cargo in Canada, so that estimate works out to only 20 fatalities per ship. The *Virginius* left Ireland with 476 passengers, of whom 158 died on board, and another 106 were ill on landing in Canada. A two-ship convoy carried 810 passengers, of which 268 died. In 1846–1847, ships bound for St. John's, New Brunswick, carried 15,000 passengers. More than 2,000 died, 800 of them on board. Irish coffin ships were named ironically: those dying on board were cast into the sea, and those who were dead upon arrival were buried in mass graves—in both cases, with no coffins.

Many of the coffin ships leaving Ireland in 1846–1847 were bound for Canada, where they emptied their diseased, dying, and dead passengers. On arrival, passengers were isolated in quarantine areas to prevent the disease causing an epidemic in their new home. The quarantine station on Grosse Isle, located near the mouth of the St. Lawrence River, was a primary arrival point for the coffin ships. The first to arrive in 1846 carried 430 people suffering from typhus. Within two weeks, 40 ships, all with typhus-infected passengers, waited on the St. Lawrence River to unload their human cargo.

Initially, the acutely ill were removed from the arriving ships, leaving the others to remain on board for a 15-day quarantine.[4] Many of those who remained on the ships were infested with lice that had fled from victims who had died, so typhus had its own captive host audience. The on-board quarantine only gave the disease more time to ravage those remaining on the ship: on one ship, nearly two-thirds of the 427 passengers who arrived at Grosse Isle were dead by the end of the quarantine period. Disembarking passengers were given a minimal examination by doctors and then released, many infected and not yet symptomatic, only to pass on the disease and die on the land they intended to call their new home.

Immigrants from the coffin ships were transferred to other cities along the St. Lawrence, each with its own quarantine area and each quickly overwhelmed by the immediate need for medical attention and the greater need to prevent an epidemic. Once on shore, passengers were kept in hastily constructed "fever sheds" that were no better than the ships they left. The huge sheds, some nearly 50 meters long, had no ventilation; some had no sanitary facilities. Over-crowded sheds contained two-level bunks with 2–3 people in each bunk, regardless of age or sex. Minimal food was provided; clean water was scarce and feverish patients had to wait hours for a small sip. Doctors and nurses attending typhus-infected patients became infected themselves, and many died.

Typhus did not discriminate among victims. Care for patients in Montreal's fever sheds was provided largely by the "Grey Nuns," members of a Catholic order. Of the 40 nuns who cared for the ill, 30 were sickened themselves, and seven died. Priests attending to the spiritual needs of the dying immigrants became sickened and died within days. Fearing spread of the epidemic disease, local mobs threatened violence. Montreal's mayor, John Easton Mills, was able to stop a riot, but he died soon after he provided care to patients.

Over the five-year period 1846–1851, nearly half of the ships with passengers fleeing Ireland landed in New York. But during the first two years—the years of the "coffin ships"—only a few went to the United States. Still, one of them carried infected passengers who initiated an epidemic in New York in 1847. Records from one hospital showed 137 newly arrived Irish patients, of which 11% died. The situation in New York was far better than in Canada. Emigration laws enacted in the United States in 1846 limited the number of passengers that could be carried on ships that were headed for US ports, and made the ships' owners responsible for the emigrants for two years after arrival. This price was too steep for most ship owners; instead, the ships headed for Canada, where no immigration laws were in place to limit the ships and their human cargoes. And there the Irish emigrants who survived the voyages and the fever sheds traveled to nearby cities to begin a new life or to move on. Every one of them must have had body lice and many of those were infected lice, and so typhus spread into the Canadian port cities, and beyond. Still, no one recognized that lice were the vectors of the disease. Despite all the efforts by doctors, nurses, nuns, and others, some 20,000 Irish immigrants died in the fever sheds and coffin ships in the first two years of the famine-driven diaspora.

LICE AND TYPHUS

Epidemics in the United States in the 1800s were not carried by Irish immigrants alone. In the first half of the 19th Century, typhus epidemics occurred in several major US cities, including New York, Baltimore, Philadelphia, and Washington, DC. Typhus also played a role in the US Civil War, although the impacts of "camp fever" on troops were less than those caused by the similarly named typhoid fever—accounts of casualties often combined victims of both diseases.

By the end of the 19th Century, no one had yet recognized the role of lice in the spread of typhus. Lice were not implicated until 1903, when Charles Nicolle, Director of the Pasteur Institute in Tunis, Tunisia, discovered the connection. He found that typhus-infected patients could transmit the disease to other patients, but not after they had been given a hot bath and change of clothes. Once he had made the association with clothing, on further investigation he discovered that the clothes of the sick were infested with lice, and he proposed lice as vectors of the disease. Nicolle tested his theory by infecting chimpanzees with typhus, later removing lice from the sick ape and transferring them to a healthy chimpanzee, previously unexposed. Within 10 days, the healthy chimp had been infected. Lice were indeed the vectors.

Nicolle repeated the experiment several times to ensure that his results were correct.[5] He was awarded a Nobel Prize in Physiology or Medicine in 1928 for his discovery and for his efforts (though unsuccessful) to develop a vaccine to prevent typhus. His selection for the prize was not without controversy. Although he uncovered the disease vector, he did not discover the bacteria causing typhus—others would make that discovery, by 1915. And he discovered no new principles: the role of insects as disease vectors (for malaria) had been discovered by Ronald Ross in 1897, and Nicolle had applied that understanding to the transmission of typhus. But Nicolle's discovery of lice as vectors—whatever the causative agent—helped prevent serious typhus outbreaks in the early 20th Century, meeting Alfred Nobel's criterion that the prize be given to those who "confer the greatest benefits on mankind."

Discovery of the bacterial causative agent of typhus was made by several scientists in the early 20th Century. Howard T. Ricketts, pathologist at the University of Chicago, had already discovered that Rocky Mountain Spotted Fever could be transmitted by the bite of a species of tick. In 1909, Ricketts traveled to Mexico City with his assistant, Russel Wilder, where they studied epidemic typhus. There, they discovered bacteria in the blood of typhus-infected victims and demonstrated that these bacteria and the disease could be transmitted to monkeys. Sadly, Ricketts died in Mexico in 1910—from typhus—while conducting his experiments. A few years later, Czech parasitologist Stanislaus von Prowazek, along with his colleague, Henrique da Rocha Lima, continued Ricketts' work, studying typhus in a prison hospital in Germany. Both were sickened by typhus during their studies and the disease killed Prowazek in 1915. Rocha Lima survived, and named the lethal bacteria for both his fellow investigators: *Rickettsia prowazekii*.

The typhus discoveries—of lice as vectors by Nicolle and of *Rickettsia prowazekii* as the disease-causing agent by Ricketts, Prowazek, and Rocha Lima—were important, but typhus epidemics still played a role in wars and their aftermath in the 20th Century, particularly World War I.

LICE IN WAR TIME

World War I began with the invasion of Serbia by Austria in August 1914. In a major upset, the Austrian army was driven back and some 20,000 Austrians taken prisoner in the early Battles of Cer and Kolubara, with more prisoners captured in subsequent battles. Most medical personnel in Serbia had been assigned to military units, leaving little medical support for Austrian prisoners, Serbian military guards, or civilians. Within three months, typhus joined the war effort on the Eastern Front, spreading rapidly among troops and prisoners. Within just one year after fighting broke out, typhus had claimed 150,000 victims, 50,000 of whom were prisoners in Serbia. Mortality reached 60–70%, a high enough level that German commanders refused to invade Serbia for fear troops would become infected and bring the disease back to Germany.

Lice were not restricted to the armies in the Eastern Front battles. On the Western Front, troops were equally infested with lice but not infected by typhus, because *Rickettsia prowazekii* was absent. Troops on both sides on the Western Front spent considerable time in the trenches and, although spared typhus infections, were infected by another louse-carried bacteria, *Rochalimaea* (formerly *Bartonella*) *quintana*, which caused trench fever. The disease was not caused by trenches, but troops who lived for long periods in trenches with little chance of sanitation or access to clean uniforms were prone to infestations of infected lice. Although not as devastating as typhus, trench fever was a problem for troops and their commanders, as the disease was debilitating, if not fatal, often recurred every few days, and full recovery required roughly two months, during which time the ill troops were out of action.

It is unclear whether troops knew that lice caused trench fever, but soldiers wanted to rid their clothes of lice, which they called "chats." When they could, soldiers removed their uniforms and either picked lice out of them or applied a candle or other flame to the seams where lice hid. Although the verb "chat" first gained use in the 16th Century, it became more popular in those settings, when soldiers, while de-lousing their uniforms, sat in small groups talking, or "chatting." On the Western Front, de-lousing stations were set up to try to rid individual soldiers of their lice infestations. As a million British soldiers were infested by lice, and trench fever was common. Long after the conflict ended, it continued to impact the health of soldiers who had returned home. Familiar British soldiers infected by trench fever include the writers, J. R. R. Tolkein (who wrote the *Lord of the Rings* trilogy), A. A. Milne (creator of the popular character Winnie-the-Pooh), and C. S. Lewis (author of *The Chronicles of Narnia*).

Lice saved the worst of their effects for the Russian front and post-War Russia. During the last two years of World War I, and through the 1917 Bolshevik Revolution, typhus killed 2.5 million Russians, both soldiers and civilians. Returning troops and refugees brought typhus home with them along with yet another disease, relapsing fever, caused by *Borrelia recurrentis*. Armistice did not stop the killing: the louse-caused Great Epidemic killed as many as five million Russians in the five years after the war's end.

Lice and typhus were not as universally severe across the battle fronts of World War II as they were in prisoner-of-war and concentration camps. In the Battle for Stalingrad, some 90,000 Axis soldiers from the German, Italian, and Hungarian armies were captured. Most of them died as prisoners, from the combined effects of malnutrition and typhus, the two often occurring together. The incidence of typhus in Nazi concentration camps was far worse, where crowding, the lack of hygiene and sanitation, and malnutrition set the stage for lice to flourish and for disease to spread. Prisoners in the camps—tens or hundreds of thousands of them—died from typhus; most were buried in mass graves or their corpses were stacked, awaiting disposition. Anne Frank and her sister, Margot, imprisoned in Bergen-Belsen camp, were among those victims of typhus, caused by bacteria carried in the gut of body lice.

One story providing a glimmer of hope amid the horror describes how a typhus epidemic was turned back in the Warsaw Ghetto in 1941–1943. More than 450,000 of the city's Jews and other "undesirables" had been forced into a small, restricted area, just slightly more than 4 square kilometers, with a density roughly 10 times that of modern New York City, which many consider overcrowded. Lice were there, too, infesting the residents, and the crowding and poor hygiene led to an outbreak of typhus. Fearing its spread to their troops, the German army surrounded the ghetto with armed guards ordered to shoot anyone trying to leave. Among those ghetto captives were hundreds of medical personnel who created a health council, to try to keep typhus from raging out of control. As the Germans' fear of contagion mostly kept soldiers out of the ghetto, the health council was able to develop a strategy that included improved cleanliness, self-isolating, and educating new medical students on preventing epidemics. The methods that were developed and employed—with no outside resources—worked to turn back the otherwise almost-certain epidemic.[6]

LESSONS LEARNED

Italy surrendered to the Allies in September 1943. But, prior to the surrender, retreating Italian troops had brought lice and typhus back with them. On arriving in Naples, they found an overcrowded city where many people lived in bomb shelters where close proximity led to the inevitable, rapid spread of typhus. Allied forces in Italy sought to control the epidemic and keep it from spreading to their own troops. An insecticide, dichlorodiphenyltrichloroethane (DDT), had recently been added to the arsenal against disease-transmitting insects, especially against malaria vectors in the South Pacific (Figure 8.2). The large-scale dusting of more than 1 million citizens of Naples with DDT powder stopped the outbreak in its tracks.

The decisive strike made against typhus is considered one of the great medical successes of World War II. In retrospect, however, the rapid deployment by one victorious power of a new chemical against a disease carried by the people of another, vanquished, country has also been called into question for ethical reasons. Still, the Allies' rapid response prevented what would have been a large-scale epidemic, that would have prolonged the war and caused massive casualties among both troops and civilians. DDT was similarly used in 1946 by Allied occupation troops and Japanese physicians to prevent a typhus epidemic from occurring in Japan after the war in the Pacific ended. In both situations, in Europe and Japan, the use of DDT in a swift strike to kill the vector—the body louse—prevented terrible typhus epidemics, and the lesson learned was that control of the insect vector, using a rapid and large-scale strike, could be successful. Post-War Europe or Japan could have suffered the same fate as Russia after World War I, with millions of deaths in another Great Epidemic.

Figure 8.2 U.S. soldier demonstrating hand-spraying equipment used to apply DDT.
Source: Courtesy Centers for Disease Control. Image 2620.

Regardless of the later indictments of DDT for long-term environmental harm, DDT used against lice (and mosquitoes) was a powerful weapon in the Allies' arsenal.

One other lesson learned came—ironically, given German atrocities in World War II—from a German physician, nearly a century earlier. Rudolf Virchow, a German physician, was hired to study the typhus epidemic in Upper Silesia (today, southern Poland) in 1847–1848. Virchow, unlike many of his contemporaries, was a keen observer and studied patients in their environment. Upper Silesia was an industrial region with ample deposits of coal and iron ore, but it was also an area beset by poverty. Virchow considered the typhus victims, who were trapped in their poverty, as he developed a plan to quell the epidemic. Success eluded him, providing only minimal relief to the victims. However, the report he authored on the epidemic contained ideas that changed the practice of medicine globally.

From his Silesian experience, Virchow developed the idea that "medicine is a social science" and that disease needed to be considered in context. Most people living in the middle of the 19th Century harbored body lice, but it would be 50 years before lice were recognized as vectors of typhus. Virchow observed that the disease mostly impacted poor patients, who lived in substandard housing, practiced minimal personal hygiene, and had little access to sanitary facilities. He recognized that successfully turning back disease, like the Silesian typhus epidemic, required more than simply treating individual patients for their symptoms; it required attacking poverty, which would enable the poor to better themselves, and this would help prevent disease. The weapons needed for the fights Virchow envisioned were democracy and education. For his radical statements and beliefs, he was relieved of his medical position. Still, he proceeded, seeking to improve the health and care of the citizens of Berlin, fighting for a sanitary sewer system, as well as separate and clean water supplies. His other lasting contributions include his co-founding of the field of anthropology and founding of the field of social medicine, with the major tenet being that diseases are more than simply biological processes run amok. Virchow's visionary leadership and social action seem remarkable when viewed today. But while visionary and progressive, at the same time he opposed Darwin's theory of evolution and rejected germ theory. A lot of contradictions can be packed into one person!

And so it was that lice—specifically body lice—the insect with its accompanying gut bacteria (*Rickettsia prowazekii*) impacted human history. Lice are now believed to have carried *Yersinia pestis*, which killed millions with bubonic plague (Chapter 4–7). Keep in mind, however, that lice did not kill anyone at all. None. Not millions, not zillions, but zero. The killers were bacteria the lice carried in their gut, which, ironically also killed the lice. Lice carried the bacteria that exacerbated the physical toll on the Irish who tried to escape the Great Hunger and killed countless soldiers and citizens throughout all of Europe in World Wars I and II, but lice and the diseases they vectored led to the practice of social medicine.

The remarkable story of lice is not over. *Pediculus humanus*, *Rickettsia prowazekii*, and typhus didn't vanish after introduction of DDT. In the first few decades of the 21st Century, people still flee and are displaced by wars. In the aftermath of disasters—both natural and caused by humans—refugees are all too often crowded together in refugee camps or disaster shelters, placing people in close proximity to one another. These situations provide the ideal conditions for that 5 mm long insect to cause the next epidemic of typhus.

NOTES

1. A number of years ago, RNW's daughter, Emily, was pre-school age and one of her favorite foods was rice. It would be a few years before Emily could clearly enunciate and differentiate between "r" and "l" consonants in her speaking. The family was in

a Mexican restaurant in San Antonio, Texas, and Emily's mother took her brother, Neal, to use the bathroom. As they were returning, the waitress brought a bowl of rice to the table. Emily, greatly excited, saw her mother across the restaurant and yelled, loudly, very loudly, "Hey, Mom! We got lice!" The announcement was no doubt unsettling to those seated in the adjacent booths and tables.

2. Infections caused by the bite of the "kissing bug," mentioned in Chapter 2, are aided by scratching the itch, resulting in insect frass and microscopic pathogens (*Trypanosoma cruzi*) being rubbed into the wound.

3. The source for potato consumption in the 1840s is: John Keating, *Irish Famine Facts* (Oak Park, Carlow, Ireland: Teagasc, 1996). The quantity and mass of potatoes consumed seems impossible, but were corroborated, albeit with slightly smaller figures, by other sources that state that males of that time consumed 8–10 lbs (3.6–4.4 kg) per day, with women consuming 6–9 lbs (2.7–4 kg) per day.

4. Chapter 5 tells of the origin and etymology of "quarantine," which was used to try to keep the plague from entering port cities in Italy in the 15th Century.

5. Repetition of experiments must not have been the institutional protocol for Pasteur Institute scientists at the turn of the 20th Century, and so Nicolle seemed to be exceptional in this regard. Chapter 6 tells of the lack of experimental repetition by Paul-Louis Simond, scientist at the Pasteur Institute laboratory in Karachi, Pakistan. Insufficient repetition in his study of Oriental rat fleas as vectors of bubonic plague bacteria led to long delay in acceptance of his findings.

6. The tales of the heroic physicians and other medical professionals trapped in the Warsaw ghetto should be required reading for everyone. Jewish physicians, working with no advanced medicines, only with what they had in the ghetto, demonstrated that their simple methods could prevent the worst of an epidemic, thus defeating the German plan to let typhus kill the ghetto's residents, in which German physicians were complicit. Unfortunately, all the efforts to quell the epidemic amounted only to a delay in this ultimate goal, because the ghetto was destroyed in 1943, with any remaining residents shipped to the Nazi camps—aided by record keeping punch cards, whose origin occurred from weaving silk, as told in Chapter 2.

SECTION 4

———⌇———

Aedes aegypti and Yellow Fever

—1418 CE   Madeira discovered by Portuguese
—1441 CE   Portuguese take first slaves from West Africa to Europe
—1452 CE   Sugarcane production begins on Madeira
—1500 CE   Cabral "discovers" Brazil
—1516 CE   First sugarcane cultivation in Brazil, using "Creole cane"
—1617 CE   Mendes de Vasconcellos starts export of 50,000 Angolan slaves
—1619 CE   Slaves from British privateer sold in Virginia Colony–first slaves in colonies
—1640 CE   Dutch take sugarcane to Barbados, establish production in Caribbean
—1647 CE   Yellow fever arrives in Americas (Barbados 1647, Yucatan 1648)
—1672 CE   British Royal Africa Company formed to begin British slave trade
—1793 CE   Yellow fever epidemic in Philadelphia, first in the new United States
—1828 CE   First yellow fever epidemic in New Orleans
—1833 CE   British pass Slavery Abolition Act, eliminating slavery in British colonies
—1845 CE   Panama Railroad Company formed
—1878 CE   Yellow fever epidemic in Memphis and southern United States kills 20,000
—1880 CE   Panama Canal project begun by French (abandoned in 1889)
—1881 CE   Carlos Finlay proposes mosquitos as vectors of yellow fever, but is ignored
—1898 CE   Spanish-American War; Henry Rose Carter finds human incubation period
—1900 CE   Walter Reed chairs Yellow Fever Commission, mosquito experiments begin
—1902 CE   Walter Reed dies of appendicitis
—1904 CE   William Crawford Gorgas begins eradication of *Aedes aegypti* in Panama
—1906 CE   President Theodore Roosevelt visits Panama Canal construction site
—1914 CE   Panama Canal opens; success due to eradicating yellow fever

CHAPTER 9

ᴄᴧᴐ

The Bridge Connecting Silkworms
to Mosquitos

History is not a dry list of events—it is a web of complex connections. Seemingly disparate actions and people converge, generating a story. How did you learn history? Hopefully, as a series of rich tales that enlivened events both significant and seemingly minor. Too often, history is taught without the underlying color, making connections between topics difficult to see. The focus of this section (Chapters 9–12) is yellow fever mosquitos (*Aedes aegypti*) and a disease they carry. In this chapter, we weave a number of seemingly disparate threads together and extend the Silk Roads west to the Americas.

Before we dive into mosquito biology or yellow fever we must address some questions: Where did yellow fever and the mosquito come from? And how did yellow fever get to the Americas? After all, yellow fever and yellow fever mosquitos are not native to the Americas, but to islands off the coast of southeastern Africa. Short answer: the trans-Atlantic slave trade. That answer may seem straightforward—Europeans brought yellow fever and yellow fever mosquitos from Africa to the Americas. Not so fast! The story is a web, weaving together seemingly disparate entities. Who would have thought Mediterranean monk seals would be linked to yellow fever? Or that yellow fever is connected to the silkworm moths? Or to the small Atlantic island of Madeira 600 years ago? Or to a volcano in Sumatra? The link is a plant—sugarcane—transported on the Silk Roads from its ancestral home in Southeast Asia. The story brings together sugar, slavery, and voyages during the Age of Discovery, which will enable us, in the following chapters, to view yellow fever with full understanding.

The Silken Thread. Robert N. Wiedenmann and J. Ray Fisher, Oxford University Press. © Oxford University Press 2021.
DOI: 10.1093/oso/9780197555583.003.0009

SUGAR

Humans have an innate drive to eat sugar. All animal bodies need glucose for energy to survive; it is required for the functioning of all cells. But we humans have an even greater need for glucose than other animals. Large brains—of a size, such as ours, that allows for solving problems, remembering past events, thinking conceptually about situations, or projecting future events—require large amounts of energy, and that energy is derived from sugar.

Sugar is a modern name given to a variety of chemical compounds. The word itself is ancient, having gone through a number of iterations. Its most recent origins are Middle-English and Anglo-French, with their words *sucre* or *sugre*. Before those, the Latin *zuccarum* was derived from the Arabic *sukkar*, which traces back to the Sanskrit *sarkara*, meaning pebble. Sanskrit, the scared language of Hinduism, arose in southeastern Asia in the Bronze Age. From this ancestral source, both the plant from which sugar is derived and its name dispersed to the rest of the world, mostly along the Silk Roads.

Sugar is basically a water-soluble carbohydrate—carbohydrates being one of the three major nutrient groups (along with proteins and fats)—and it occurs naturally in plants and in milk produced by mammals. Sugars are easily identified in lists of food ingredients by the suffix "-ose": glucose, fructose, lactose, sucrose, among others, are all sugars. Glucose and fructose, found in fruits and honey, are simple sugars, or monosaccharides. Sucrose, the table sugar you add to your coffee or tea, is a disaccharide formed by linking glucose and fructose.

The human drive for sugar exists to feed energy-demanding brains and is enabled by having taste receptors on our tongues that are specific to sugars. The two go together: no demand, no taste receptors; no taste receptors, no way to satisfy the demand. As a result, we *crave* sugar. And that craving has gone hand in hand with increasing brain size throughout our evolution. This may seem simple, but the full story is complex and fascinating, and it is an integral part of a larger story of symbiosis and human evolution. That larger story would be best told in another book. For now, and in this book, we have sugar and the demand for it.

SUGARCANE

Supporting the human drive for energy is a huge sugar industry and infrastructure. In 2018–2019, the world produced nearly 180 million metric tons of sugar and 80% of that from sugarcane (the remainder is from sugar beets). Sugarcane is grown in tropical and subtropical regions (whereas sugar beets are grown in temperate regions) in more than 120 countries, but with India, Brazil, Thailand, and China producing most of the world's sugar. In terms of overall biomass produced, sugarcane ranks as the number one crop plant worldwide, and production in recent years exceeded 1.8 billion metric tons. The plant consists of about 12–16% sugar, a similar percentage of fiber, and the balance—about two thirds overall—is water. In addition to being used for food, sugarcane is also a source of bioenergy, especially in Brazil, where ethanol produced from sugarcane turns gasoline into an alternative fuel.

So, what exactly is sugarcane? And where did it originate? Teasing apart history to answer those questions has proven difficult, and modern techniques are reshaping old ideas. The answer to the seemingly simple question "What is sugarcane?" is not entirely resolved. If we had asked the question as recently as only a decade ago, the answer would have been far simpler. And the answer may change again in the future. And that's OK. After all, truth is complex and elusive, and the path to it is constantly shifting. That's how science works.

The common name "sugarcane" applies to a collection of grasses that are cultivated in tropical areas worldwide. These are huge grasses, some growing up to six meters (20 feet) tall. Although they are called "canes," they are unrelated to true canes (e.g., *Arundinaria*) and, instead, are more closely related to corn and sorghum. Sugarcanes (*Saccharum*) diverged from other grasses about 1.5 million years ago. There are widely differing opinions on how many species of *Saccharum* there are. Historically, the number was about 40 species. But recent studies suggest most of those species are more closely related to other grasses. In fact, only a handful probably truly belong together as true sugarcanes. For example, *Saccharum robustum* is found only in New Guinea; *S. officinarum* is the plant that has been cultivated most widely worldwide; and *S. cultum* is a newly described species (2016) that has also been cultivated world-wide.[1] Various other species that are commonly referred to—such as *S. sinensis* in China and *S. barberi* in India—are actually just varieties of *S. officinarum*. Most interestingly, just as silkworms evolved from wild silk moths it is suspected that all these species are human-associated forms that stem from wild sugarcane (*S. spontaneum*), a species distributed from New Guinea to North Africa. Sweet mutualisms.

This new conception of the ancestral stock for sugarcane shakes up a traditional notion about its place of origin. Historically, *Saccharum robustum* has been viewed as the ancestral species and thus the island of New Guinea as the center of origin for modern sugarcane. But these recent studies change the historical view. In particular, research at the South African Sugarcane Research Institute (SASRI) synthesizes a number of seemingly disparate entities from a variety of disciplines (geophysics, climatology, biogeography, mathematics, etc.) to demonstrate the regional effects of a volcano on plant life, particularly as it related to the origin of sugarcane.

Volcano and Tsunami

The magnitude of the massive eruption of the Toba volcano on the island of Sumatra 74,000 years ago is difficult to comprehend. The caldera left by the eruption is about 100 km by 30 km in size and it is considered a "super-eruption" because of the amount of solid material (tephra) ejected. It is estimated that the Toba eruption produced 5,300 cubic kilometers of what is known as dense rock equivalent.[2] For comparison, the dense rock equivalent from the 1980 eruption of Mt. St. Helens in the northwest United States was estimated at 0.25 cubic kilometers.

The Toba volcano matters because it addresses the origin of *Saccharum robustum,* the historically accepted progenitor of modern sugarcane, in New Guinea. Ejected matter from the Toba eruption was deposited in the narrow Strait of Malacca (separating Sumatra from Malaysia) generating a tsunami. SASRI researchers calculated the magnitude of this volcano-caused tsunami and its effects on New Guinea, more than 3,300 kilometers away. They show that New Guinea was inundated to a depth of about 750 meters. Sugarcane, which requires high temperatures, would have been growing in these flooded, low elevations .

The larger analysis conducted by SASRI researchers also included studies of glaciers and sugarcane biogeography. The cooler temperatures that followed the eruption would have made it impossible for sugarcane to grow above the level of the flood on New Guinea. We're not done. The researchers then examined local languages (the eastern half of the island, Papua New Guinea, has over 800), the presence of domesticated pigs, and the various recent uses of sugarcane by New Guinea farmers. And importantly, they also examined sugarcane DNA to decipher evolutionary trends. The conclusion, derived from integrating those studies, is that New Guinea could not have been the ancestral home of sugarcane. Furthermore, *Saccharum robustum,* once thought to be the ancestor of modern sugarcanes, is likely just another variety of *S. spontaneum* that was introduced to New Guinea in modern times, along with domesticated pigs. It now appears that southern Asia (range of *S. spontaneum*) is the ancestral home of modern sugarcane.

Do these revised interpretations change how sugar tastes in your morning coffee? Yes! Add to your experience all the vibrancy of these stories and we assure you, your sugar will taste sweeter. All the mysteries, contentions, and re-interpretations of results help us to peer into the process of science, where seemingly disparate lines of inquiry are gathered and synthesized to answer questions. The results also help illustrate the silken thread that connects sugarcane to yellow fever.

That silken thread is tangled, but the knot will be loosened by the focus on sugarcane's origin and the reconstruction of the early paths along which it moved. Completely untangling the knotted history of sugarcane is not the purpose here. We pick up the thread with sugar's history, 2,500 years ago.

FROM ANCIENT ORIGINS

The sugar in a sugarcane stalk breaks down soon after the stalk is cut, and so the ancient consumption of its sugar was by chewing recently cut plant pieces, before the sweet taste was lost. That way of consuming sugar is still practiced in many parts of the world. For sugar to be useful beyond immediate consumption, it has to be processed, which requires crushing the stalks, extracting the juice, boiling it, and finally cooling it to form sugar crystals. The harvested crystals are stable and maintain all sugar's properties. The first record of sugar processing comes from India. By 500 BCE, at least, technology, such as animal-powered mills and associated tools, was used to extract juice from the sugarcane stalks as just described. Although materials and technology have been greatly improved and modernized, the process of crushing and extraction from cut stalks remains largely unchanged today (Figure 9.1).

The extraction process made crystalline sugar a desired commodity, known well beyond India. In the mid-600s CE, Emperor Taizong, the second emperor of the Tang Dynasty, apparently became interested in India's sugar production. Indian envoys brought two sugar makers to the Tang Emperor and, by the end of that century, the Tang Dynasty had its own sugar industry. Getting sugar production to the East (China) required two emissaries, one trip, and a few years. In contrast, getting sugar production to the West was a stepwise process, with multiple stops, multiple players, over nearly a millennium. For our purposes, we are concerned with only one of the many varieties that resulted from crossing *S. officinarum* with *S. spontaneum*. This important variety was distributed and grown for more than a millennium and dominated the sugar industry prior to the 1780s. It became known as "Creole cane" in the Americas after it was replaced by other hybrids from Polynesia and Java that were introduced during the 1800s. But the rest of this story is about Creole cane.

The first attempt to move sugarcane west was after Alexander the Great's conquest of the Indus Valley, but little came of that attempt. Instead, sugarcane and sugar production from India became firmly established in Persia much later, by about 600 CE (hereafter all dates are Common Era unless specifically mentioned). With the rise of the Arab Empire in 651, sugarcane became one of the crops of the Arab Agricultural Revolution, and the plant was carried throughout the then-extensive Arab world. Within 50 years, sugarcane was grown in the eastern Mediterranean and North Africa. Those regions of the Arab world were arid, and sugarcane is a plant

Figure 9.1 Cut sugarcane stalks.
Source: Regreto/Shutterstock.com

that requires water. Consequently, irrigation systems were constructed to grow sugarcane effectively. Sugar grown in the eastern Mediterranean was encountered by the Crusaders in 1099, who then took the knowledge of sugarcane back to Europe. They tried to establish sugarcane in England and northern Europe, only to find that the plant would not grow in the temperate (no, northern Europe was mostly colder than temperate) climate. After the 11th Century, processed sugar was funneled into Europe by Venetian traders through routes that passed from South Asia to Central Asia, to Western Asia, the Mediterranean, and into Europe—almost perfectly overlaying some of the long-standing routes of the Silk Roads.

In an effort to reduce the costs of moving sugar over long distances, Venetians also started to grow sugarcane themselves. Realizing how labor-intensive this process was, they brought slaves from areas around the Black Sea to work in its production. At that point, sugar production was largely concentrated in Venetian Cyprus and, later, Sicily, and the trade in slaves was on a small scale. For the next few centuries, granulated sugar was very expensive throughout Europe, because of the cost of transportation, its perceived value, and the near monopoly held by Venetian sugar merchants. What choice did consumers have? If only another power, one other than the Venetians, produced sugar! Portugal and Henry the Navigator enter the stage at this time.

Time for a brief pause. Does all this sugar history have something to do with yellow fever? Portugal? Henry the Navigator? Yes, yes, and yes. We are getting there.

PORTUGAL, PRINCE HENRY, AND MADEIRA

It was in 1394 that the fourth child of King John I of Portugal was born. Infante Dom Henrique of Portugal, later known as Henry the Navigator, was as influential in world history as anyone in the 1400s. Patron of explorers, Prince Henry was the moving force behind the Age of Discovery, through not only his support of voyagers but also his involvement with science and technology. Despite his nickname (which was given long after his death), Henry the Navigator was a land-based leader rather than a leader at the helm of ocean-going ships, but at an early age, he had been intrigued by exploration and, particularly, by Africa.

By age 25, Henry was made governor of the Portuguese province of Algarve, on the Atlantic coast, a position that aligned with his love of the sea and exploration. As the sponsor and supporter of oceanic exploration, Henry was also one of the instigators behind development of a new kind of sailing vessel. This was the caravel, a light, fast ship and an improvement on the slow and lumbering ships of the day. Part of their ability came from the sail they used: the lateen, a triangular sail first developed by Arabic sailors who plied the Arabian Sea in shallow-draft dhows. The lateen allowed caravels to sail into the wind, which afforded them greater speed and maneuverability. The lateen, coupled with the trade winds, enabled the transoceanic movement of sailing ships, and their shallow draft meant caravels could hug the coast and travel upstream into river channels. The caravel and the trade winds figure prominently later in this story.

In 1418, before Prince Henry became governor, he sent ships on an exploratory voyage on the seas of the eastern Atlantic to discover new lands. After being blown off course by a storm, the captains found shelter in the lee of an island, which they named Porto Santo (Holy Harbor). The island, 500 kilometers west of Morocco, was visited on a return voyage in 1419, when Porto Santo was claimed for Portugal. The explorers also noted a larger island on the horizon and they named the islands, and its main island, Madeira, meaning "wood" in Portuguese. Madeira is an archipelago of islands, part of the region of seamount islands known as Macaronesia, which also includes the Azores, Canary Islands, and the Cape Verde Islands.

The Madeira islands were revisited on Henry's orders in 1420, with the intent to establish a colony there. The three small islands known as Desertas—aptly named because they are "deserted" other than by a dozen or so bird species—could not support a colony. A small colony was established on Porto Santo, but the main interest was in the much larger island of Madeira. The main island is more than 740 km^2 in size and is ruggedly mountainous, with peaks reaching 1,862 meters (Figure 9.2). It had also been appropriately named: it was found by the intended colonists to be completely covered in impenetrable forest. To gain a toehold on the island, the colonists set this forest alight, starting a fire so intense that they had to take shelter in the ocean for two full days and nights, and it continued to burn for more than seven years.

Through the fire, Madeira was cleared and the land heavily enriched with a layer of ash. Where possible, the mountainous terrain was terraced, at great physical cost, to plant crops. Henry wanted to establish agriculture, to export food to Portugal and

Figure 9.2 Madeira Island terrain. Terraces used to level ground for cultivation are visible.
Source: Gerd Eichmann, CC BY 4.0

make the venture profitable, so wheat was planted. Within a few years, as wheat yields dropped off, he ordered the planting of sugarcane, believing the crop could prove profitable there. Sugarcane needs water—a lot of it—for optimal growth, and so its cultivation on Madeira needed irrigation, and irrigation required level ground, which meant further terracing. Creating terraces and irrigation channels required more labor than the colonists could provide. The Canary Islands were raided to provide slaves to dig irrigation channels through rock, but soon even that supply of labor proved insufficient—and the dangerous work depleted the numbers of slaves daily. Sugarcane production became the economic engine for Madeira, expanded, and soon even required more labor than could be had from the slaves taken from the Canaries.

INFAMY

Today, Madeira is famous for its wine. But it also is infamous, for two reasons. First, an ecological and economic disaster was brought on by the worst excesses of capitalism: an example of natural resources depleted so quickly, the pendulum swung so far, that full environmental recovery has not yet happened six centuries later. Second: the island's role in the worst of inhumane treatment, the culture of slavery. Just as the island's ecological resources were decimated in their home range, so, too, were Africans on the mainland decimated in their home range. African people were forcibly torn from their homes and families, shipped across the ocean to be treated as if they were disposable, to be used until their vitality was sapped, their lives taken. And Madeira played a major role in that story.

These two stains on Madeira's history happened mostly simultaneously. We'll start with the ecological disaster. The history of large-scale production of what came to be called Creole cane on Madeira can be dated to 1452, when the first water-powered sugar mill began operating. Within two years sugar production reached 76 metric tons, then 254 metric tons by 1472, and then at least 2,000 metric tons each year from 1506 to 1509. By that time, sugar production on Madeira exceeded that of Cyprus, the top Mediterranean producer. Sugar flowed more freely throughout Europe than it had ever done before, adding to Portugal's riches. Within 25 years, however, production failed massively and output dropped to 1472 levels.

What happened? At the beginning of the 16th Century, 10,000 hectares of sugarcane were planted annually, while, at the same time, 500 hectares of forest were razed to fuel the sugar boilers. Processing sugar used a lot of energy, in the form of wood burned to fuel the boilers. One estimate is that 50 kilograms of wood was needed to make 1 kilogram of sugar. Fifty to one. Think about that as you spoon sugar into your coffee—although, modern processes are more efficient, they are still energy hungry.[3] On Madeira, deforestation, undertaken on a large scale to feed the furnaces to boil the sugar syrup, had moved beyond the immediate reach of the sugar mills and into the lower mountains—the forests of the higher elevations were out of reach. Islanders had stripped half of Madeira's accessible forest by 1509. Think about that. Half of the accessible forest—160 square kilometers—was gone in less than 60 years, and it did not come back. Ironically, after wine had replaced sugar as a valuable export, even the wood needed to build casks for the island's famed wine had to be imported from across the Atlantic. That's the ecological disaster.

The second disaster was worse. We begin in the year 1441, the year that changed everything. Two explorers were sent by Prince Henry on new caravels to explore the west coast of Africa, which, at that point was known only as far south as Cape Bojador, in modern Western Sahara. As they had different objectives, the two explorers, Antão Gonçalves and Nuno Tristão, traveled separately and planned to meet up later. Tristão was a true explorer, sailing to discover new lands and return to Portugal with gold. Gonçalves was sent to hunt Mediterranean monk seals (*Monachus monachus*) and return with seal skins. He found seals on the shore of Cabo Blanco, at the border of modern Mauritania and Western Sahara, and he filled his boat with seal skins. It is not known how numerous the seals were in 1441, but today they are the rarest of all seals, with 600–800 surviving, almost all in just two places—the Aegean Sea and the Atlantic coast at Cabo Blanco, though a few survive on the Desertas islands of Madeira. Interestingly, most of the remaining seals are found on Cabo Blanco, the same beaches where Gonçalves hunted them in 1441.

THE SLAVE TRADE BEGINS

Before meeting Tristão near Cabo Blanco, Gonçalves and his crew "obtained" a Berber local and a black slave—under his own initiative, not on Prince Henry's orders. Accounts differ; some tales are generous, stating he "bought" the two, others that he attacked and captured them. Joined by Tristão, the two captains then obtained ten more slaves—again, whether "purchased" or taken is up to interpretation. One of those taken was a chief of the Azenegue Berbers. Gonçalves returned with the captives to Portugal while Tristão continued to sail further south. Perhaps Gonçalves thought he would get rich by his actions. The following year, he returned to West Africa and brought the captured Azenegue chief with him, hoping to trade the chief for a large number of Berber slaves. What he received in trade for the chief were ten slaves, a small amount of gold powder, and a large number of ostrich eggs. Yes, ostrich eggs. A lot of them. Gonçalves returned to Portugal, no richer, save for the 10 slaves. And the ostrich eggs. By his action, seemingly inauspicious in its scope and magnitude—a simple voyage to hunt seals—Gonçalves initiated what would become the Portuguese slave trade from western Africa to the Americas.

Gonçalves may have been the first to take black West African slaves to Europe, but Tristão perpetuated the transport of captive slaves to Portugal. On two more trips, he returned to Portugal with fewer than two dozen captives, but he also passed beyond the southern edge of the Sahara, claiming to have reached the coast of present-day Senegal. His reports on how easy it was to capture African slaves inspired many other captains to voyage to the area to take slaves, though they found neither the large number nor the easy captures they had anticipated. Tristão was ultimately killed by his intended captives. Still, the Portuguese found themselves in the slave-transporting business. More than "found themselves" in the business, they embraced it and became slave traders of the highest order. Exploitation of the newly found source of labor mushroomed and Madeira, as a destination for slave laborers, was in the middle of it all.

With added labor, Madeiran sugar production peaked in 1509, but then dropped by 75% within 20 years, even as slave numbers increased. The production increase from 1489 to 1509 masked the symptoms of the coming crash. Large-scale monoculture has its issues, and the Madeira sugarcane crop was not immune to them. After a yield boost as the land was converted from forest, especially with the addition to the soil of ash from burned trees, depletion of soil nutrients began. Crop monocultures are prone to invasions by adventitious weeds, which get a foothold and then grow exponentially, and significant labor was needed to battle the yield-robbing invaders in Madeiran sugarcane fields. Then insect pests arrived, first noted in a 1502 report as "caterpillars." Worldwide, sugarcane hosts a large number of pests, especially moths, whose caterpillars capitalize on the rich source of energy the sugarcane provides. Although the identity of the caterpillars of 1502 is not certain, the culprits were likely either armyworms (whose feeding strips entire plants) or stemborers (whose larvae tunnel inside plant stems, reducing sugar content and stunting or killing plants). Then came the rats, pests that still invade sugarcane fields worldwide.

The combination of all the simultaneous stresses to Madeiran sugarcane—nutrient losses, weeds, insect pests, and rats—conspired to reduce yields, except for one externality: slaves. Although perhaps only 200–300 slaves were shipped annually to Madeira in the early years, the island's slave population continually increased, reaching more than 3,700 by 1529. Throwing more slaves at a doomed crop only delayed sugarcane's eventual collapse. Productivity of slaves, measured as the amount of sugarcane produced per slave, peaked with the greatest crop yields in 1509. However, productivity declined by 75% in the next two decades, and sugarcane yields declined despite the sharp increase in the numbers of slaves.[4]

SUGAR—AND SLAVES—ON THE MOVE

All the while, European demand for sugar was increasing beyond the potential maximum production on Madeira. The Portuguese introduced sugarcane to the Atlantic islands of the Azores and Cape Verde, then to São Tomé, far to the south. The Azores, located further north, proved to be too temperate for the warmth-loving sugarcane. And the Cape Verde Islands, discovered in 1455 about 500 kilometers off the coast of Senegal, proved to be too dry for the water-loving sugarcane. Like the tale of Goldilocks, São Tomé proved to be just right.

The equatorial island of São Tomé sits about 250 km off the coast of Gabon. The larger of two islands (Príncipe is the other), São Tomé, at about 840 km² in size, offered tropical temperatures, abundant rain, rich soils, and flat lands. The Portuguese colonial push and their desire for sugar converged on São Tomé. Colonization of the islands in 1493 was led by Alvaro de Caminho, whose charter from the king included producing sugarcane and allowed him to take, as colonists, condemned prisoners and, curiously—but tragically—some 2,000 Jewish children, aged 2–10, wrested from their parents. His charter also permitted him to take slaves from the African mainland. Within two years, sugar from São Tomé, produced by slave labor, was being sold in Europe.

At this point, the Atlantic trade winds came into play. The trade winds form a large circular pattern, called the Volta do Mar, or "turn of the sea," with winds carrying ships westward near the equator, then northwest, turning back to the northeast in the North Atlantic and then southeast in a return to Africa. Portuguese sailors understood and used the predictable winds to aid in their travels. The same trade winds had carried Columbus to Hispaniola in 1492. In 1500, Pedro Álvares Cabral was sent by King Manuel I with 13 ships to sail around Africa to Asia, to develop a Portuguese trade presence. Cabral's ships caught the trade winds and landed on the shores of Brazil. Thinking Brazil was a large island, he claimed it for Portugal, even though it had been discovered previously. Cabral sent one ship back to Portugal to notify the king, and he continued on to India.

Initially, Brazil failed to impress the Portuguese explorer. Cabral's scribe, in his report to King Manuel, wrote that the natives had little to offer of any value—no spices, no gold, no jewels. For nearly three decades, Brazil was a colony in name only, mostly ignored by Portugal—until the French showed an interest. After territorial skirmishes with the French, the Portuguese became more serious about the colony. Portuguese settlers, many of whom were from São Tomé and Madeira, arrived and, by 1516, were cultivating Creole cane. Unlike the islands from which the colonists and the Creole cane were brought, northeastern Brazil was broad and flat, and ample rainfall meant no irrigation was needed. Equally important, the vast forests yielded plenty of wood for processing sugar. As Brazilian production scaled up, so the need for labor once again reared its head. Rounding up and capturing local natives was attempted, but the enslaved natives proved to be inefficient workers. Instead, the model perfected by the Portuguese in Madeira, Cape Verde, and São Tomé was tried, using captured African natives to provide the necessary labor.

STEALING A PEOPLE

The Portuguese colony of Angola was established in 1575, with a charter given to Paulo Dias de Novais. The goal of colonization was to establish trade, and that trade was to be in slaves. The king's charter explicitly included subjugation of people from the local kingdom of Ndongo. At first, the Portuguese bought slaves, but soon realized that a far more profitable venture would be to take slaves, rather than pay for them. To acquire enough slaves, Dias de Novais pushed inland more than 200 km. He recognized that he did not have enough soldiers to capture all the slaves required, so they set about pitting one local kingdom against another—the captured people from a vanquished kingdom then became available for purchase.

When Luis Mendes de Vasconcellos became governor of the colony of Angola in 1617, he made enslavement an even greater priority. Between buying the spoils of war and capturing thousands of others, the Portuguese processed a continuous supply of humans. Between 1617 and 1621, de Vasconcellos's minions forcibly captured 50,000 Angolans[5] and moved them to the port city of Luanda, where they were loaded onto ships and sent to Brazil. Fifty thousand individual human lives stolen in three years! Over the 240 years of the colony's slave trade, more than a million people from southwestern Africa were sent to the Americas, beginning with Brazil. But 50,000 slaves in just three years was unmatched in the rest of the colony's inhumane history of stealing a people.

Enslavement of people did not begin with the forced movement of Africans to Brazil by the Portuguese. There was a long and widespread history of slaves being used in many societies—notably by the early Greeks and Romans in the first millennium BCE, but also elsewhere, in Africa and in South America. For centuries, wars resulted in the capture and sale of vanquished peoples in exchange for desired goods, and the wars among empires along the Silk Roads resulted in the enslavement of the losing tribe or empire. Even in the absence of war, stealing people from nearby lands was a regular practice in early Asia and Eastern Europe. The word "slave" comes from the name of the people, the Slavs, who were often sold, traded or stolen, often by other Slavs. The merchants who perfected the process of trading humans from Africa were Arabs. Their incursions into the Horn of Africa and across the savannas of the continent began in the 7th Century and resulted in some 10 million Africans being shipped to the Arabian peninsula and beyond. By the beginning of the trans-Atlantic movement of African slaves in the 16th Century, the buying and selling of African people as goods had been ongoing for millennia.

But everything changed in the 16th Century as European powers became slave traders, especially the Portuguese. First, the scale of the venture, capturing and shipping slaves, was nearly industrial in size and scope. Second, the trade was ubiquitous, with all the major European empires of the 16th Century participating. The Portuguese may have begun the trade, but they were not alone in taking slaves to the Americas. All slave taking is terrible, but the horror of Portuguese slave taking came from their indifference to the slaves' well-being and the lack of any value being placed on the slaves or their lives. Slaves were treated as a commodity to be exploited fully—much as the resources on Madeira had been exploited fully. The mortality level associated with Portuguese slave trading was unlike that associated with earlier slave traders, such as the Romans or Egyptians, who valued and cared for their slaves, many of whom remained with families for years. In sharp contrast, Portuguese slave traders de-valued slaves and accorded them little to no care or concern. Those slaves who died in Portuguese hands were simply replaced by more.

The convergence of Portuguese slave-trading, the colonization of Brazil, and the increase in sugar production was a cycle with a positive feedback—an increase in one factor caused increases in the others. Early in the 17th Century, shipments of slaves to Brazil averaged about 10,000 humans shipped per year to provide the plantations with a steady supply of labor (Figure 9.3). The demand for slaves was due not so much to increases in the amount of land dedicated to sugarcane production but to the brutal working conditions. The very things that made Brazil appealing for growing sugarcane made slaves' lives even tougher. The heat and humidity were stifling, and the slaves worked from before dawn to after dark. Great forests were cleared and trees felled to open up land and to fuel sugar boilers; work that was both physically demanding and dangerous. Other conditions were equally challenging. Mosquitos and other insect pests bit or stung the slaves in sugarcane fields, small scorpions lived

Figure 9.3 Slavery in Africa.
The Treaty, vintage engraved illustration. Journal des Voyage, Travel Journal, (1880–81).
Source: Morphart Creation/Shutterstock.com

in the leaf sheaths of the plants, and rats in the fields were followed by large numbers of fer-de-lance. The extremely venomous snakes preyed on the rats and their deadly bite was a constant concern[6] to field workers. Survival of slaves for even a few years was uncommon, and so the need for the continuous stream of new slave workers.

VOYAGES, PERILS, AND HISTORY

English privateers made their living by capturing ships flying the flag of Spain (or Portugal, as the two countries were united under one monarch at that time). But when England signed a truce with Spain, English privateers had to go to another country to get letters of "Marque"—which gave them the freedom to capture ships of their enemies. One connected tale relevant to our history is of two privateer ships: the *White Lion*, captained by John Jope, and the *Treasurer*, captained by Daniel Elfrith. The two privateers, carrying letters of Marque from Holland and Italy, respectively, sailed in 1619 intending to capture the gold and other cargo on a Spanish ship. At the same time, the Portuguese slave ship, the *São João Bautista*, had departed from Angola with 350 slaves, bound for the Yucatan peninsula of Mexico. The rough voyage across the Atlantic proved deadly for 120 of the slaves—a 30% mortality rate that was apparently typical. After unloading some of the surviving slaves in Jamaica, the *São João Bautista* continued toward its destination. But the Portuguese ship was found and captured by the two privateers off the coast of the Yucatan.

The two captains were unable to take all the slaves, so they split the cargo on board and divided the 60 slaves—the *White Lion* taking twenty, the *Treasurer* forty and headed for the British colonies of North America. The *White Lion* arrived first, on August 20, at Old Point Comfort in the Virginia colony, just north of latter-day Norfolk, and the 20 slaves were traded for supplies—corn. When the *Treasurer* arrived, its captain failed to get a buyer for his human cargo, so he sailed to Bermuda, where the slaves were sold to work on the sugar plantations there. The 20 slaves captured in Angola, who survived the Atlantic crossing, only to be stolen by privateers and traded for corn in the Virginia colony? They were the first slaves in what would become the United States, and they are the subject of the ongoing 1619 Project.[7]

Slave ships continued to bring their human cargo across the Atlantic. Sugarcane production expanded into the Caribbean, first by the Dutch, then augmented by Spain, Britain, and France, the other global powers with colonies in the Caribbean at that time. As the 17th Century progressed, the point of origin of many of the slave shipments shifted north, to the coast along present-day Gabon and Cameroon (Figure 9.4). Somewhere in the movement into more tropical areas, before the middle of that century, slave ships picked up new passengers—yellow fever mosquitos (*Aedes aegypti*) and their traveling companions, yellow fever virus. One of the slave ships that escaped the dreaded English privateers landed on the Yucatan peninsula of Mexico, where it unloaded all its passengers—including mosquitos—producing, in 1648, the first recorded yellow fever epidemic.[8]

As detailed in Chapters 10 and 11, the intricacies of the host-mosquito-virus cycle made that epidemic unlikely. To understand just how unlikely, we need to provide a little information about the virus and mosquito (both will be explained more fully in Chapters 10 and 11). Neither *Aedes aegypti* nor the yellow fever virus were found in the Americas before European colonization; they came on slave ships. Both had ancestral homes in southeastern Africa and made their way to West Africa, through means and routes that are not yet understood well. A series of factors had to align for an epidemic to occur. The mosquito and virus needed to occur near the Slave Coast and its ports. At least one slave needed to be infected by the virus. Both the mosquito and virus had to board a ship from a Slave Coast port. The virus would have been transported in an infected slave. Mosquitos had to board, either as adults, which can live for 30 days, or as eggs laid in barrels of fresh water; the eggs would have been viable for the voyage's duration. The virus had to be picked up from an infected human by a mosquito, and the mosquito had to bite another human at the appropriate time. Infected humans and surviving mosquitos were ultimately unloaded on land. Every factor had to be in place, making an epidemic's occurrence unlikely, and most voyages did not result in an epidemic. However, rare events, when given enough opportunities, actually do happen. The sheer number of voyages of slave ships created the opportunity for the unlikely yellow fever epidemic. However, it is unlikely that enough other factors would have been sufficient for *Aedes aegypti* and yellow fever virus to have become established in the Americas in the absence of the slave trade.

This thread began as a group of grasses that diverged from other grasses 1.5 million years ago, followed a path along the Silk Roads out of southern and southwestern

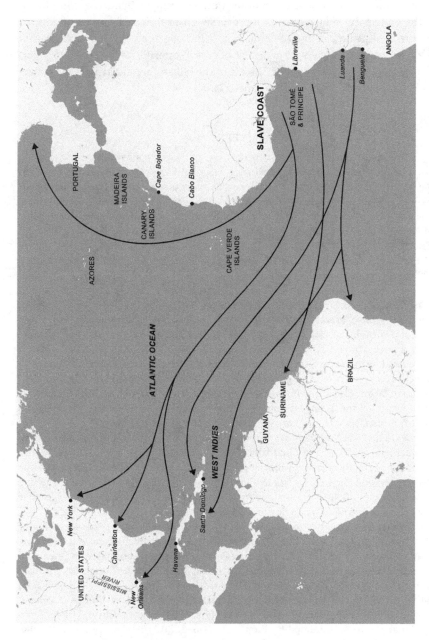

Figure 9.4 Major trans-Atlantic slave routes; from Angola and the Slave Coast to western terminuses in Brazil, the West Indies, and the United States.

Source: Sarah Silva, MapShop.com

Asia, then passed through the Mediterranean and on to the Eastern Atlantic Ocean, to Madeira, and São Tomé. The silken thread and its complex connections reached across the Atlantic, stretching from Angola to Brazil, to the Caribbean colonies, and to the Yucatan. The plantation system of sugarcane developed in northeastern Brazil began the systematic subjugation of a broad swath of different groups of African people, eventually numbering in the millions. What began on the Silk Roads, based on the products from the salivary glands of silk moth caterpillars, morphed to sugar, slaves, mosquitos, and yellow fever. The arc of the silken thread changed the trajectory of history in both Africa and the Americas.

NOTES

1. For a broad-scale analysis of the history, biology and identity of sugarcane, pointing out that what has been called "sugarcane" is not quite so simple, based largely on molecular analyses and their interpretations, but also including studies from numerous other disciplines, see D. L. Evans, "A Geophysical and Climatological Assessment of New Guinea—Implications for the Origins of *Saccharum*," *BioRxiv* (2020), https://doi.org/10.1101/2020.06.20.162842; D. L. Evans and S. V. Joshi, "On the Validity of the *Saccharum* Complex *And The* Saccharinae Subtribe: A Re-assessment," *BioRxiv* (2020), https://doi.org/10.1101/2020.07.29.22675; D. L. Evans and S. V. Joshi, "Origins, History and Molecular Characterization of Creole Cane," *BioRxiv* (2020), https://doi.org/10.1101/2020.07.29.226852.

2. The 2014 geological and vulcanological study by Costa and colleagues is a fascinating read and offers insight into just how powerful the eruption by Toba was. For example, their model estimated that the mass eruption rate was 175 million metric tons *per second* and, as in the text, the amount of material released was 5,300 cubic kilometers of dense rock equivalents, or greater than 21,000 times more powerful than the 1980 eruption of Mt. St. Helens in the northwestern US. A. Costa, et al., "The Magnitude and Impact of the Youngest Toba Tuff Super-Eruption," *Frontiers in Earth Science* (2014) https://www.frontiersin.org/article/10.3389/feart.2014.00016.

3. In the 1990s, RNW worked on biological control of sugarcane pests (stemboring Lepidoptera) and had the opportunity to visit several sugar mills. The large mill operated by the Rio Grande Valley Sugar Growers, near Santa Rosa, Texas, could be detected several kilometers away by the vibrations in the ground from the processing (crushing) of 11,000 metric tons of sugarcane per day, to produce 160,000 metric tons of sugar and 60,000 metric tons of molasses annually.

4. For an excellent, two-part history of Madeira and its ecological disaster, telling of the convergence of all the factors leading to the inescapable collapse of sugarcane in Madeira, which led to the initiation of Portuguese sugar production in Brazil and slavery to support it, see: J. W. Moore, "Madeira, Sugar, and the Conquest of Nature in the 'First' Sixteenth Century: Part I: From Island of Timber to Sugar Revolution, 1420–1506," *Review* (Fernand Braudel Center) 32 (4 2009); J. W. Moore, "Madeira, Sugar, and the Conquest of Nature in the 'First' Sixteenth Century: Part II: From Regional Crisis to Commodity Frontier, 1506–1530," *Review* (Fernand Braudel Center) 33(1 2010).

5. The kingdom of Angola extended farther north than the northern boundary of the current country; the kingdom included land from the Okavango River on the south, extending north to include current Gabon.

6. The first sugarcane field RNW visited in Honduras held surprises. Seemingly, every leaf sheath pulled open contained a small scorpion. Anxious to jump into the field and search for stemborer larvae, RNW was warned by three students from the Pan American School of Agriculture (Zamorano) in eastern Honduras to "watch for 'yellowbeard'"—the English translation of a local name for fer-de-lance. The warning led to an immediate exit from the field (without seeing the snake). Further sampling resumed along the outside edge of the field.

7. The 1619 Project is an effort begun by the *New York Times* on the 400th anniversary of the arrival of the first slaves in the colonies in what became the United States. The project chronicles the time from 1619, when the first slaves were put ashore and sold in the Virginia colony, to the present, illustrating the history of slavery and the contributions of slaves and their descendants to the history and development of the United States: https://www.nytimes.com/interactive/2019/08/14/magazine/1619-america-slavery.html .

8. There was a yellow fever outbreak on Barbados in 1647, though not reaching epidemic levels.

CHAPTER 10

༺༻

Yellow Fever in the United States

Put simply, yellow fever is a dreadful disease. Unlucky souls can die within a week of becoming infected, after bleeding from their eyes, vomiting blood, and succumbing to kidney failure and delirium. The pathogen is transmitted by mosquitos, who don't care if you wash your hands, live in clean communities, or have social standing. Effective vaccines exist, but the disease still rages. One study estimated that 30,000 people die globally out of the 200,000 infected annually, another that in 2013 alone as many as 60,000 died from yellow fever, mostly in parts of Africa where vaccines, relevant education, and aid are all scarce. But as jaw-dropping and heart-breaking as the current state of yellow fever may be, only a few generations ago, it was even worse. Even for those safe in Western culture who know its history, the name yellow fever itself still inspires dread. The disease may conjure images of faraway, steamy jungles, rife with all things deadly, but yellow fever put its mark on Western culture too. Although yellow fever originated in East Africa, and still wreaks havoc there, it didn't stay there. Yellow fever made its way to the Americas, and it did so by travelling with the slave trade.

The trans-Atlantic movement of slaves from east to west was just one side of a triangular trade route (Figure 10.1). Ships embarking on this three-part voyage took manufactured goods (e.g., weapons) from western Europe to trade on Africa's Slave Coast ("First Passage"), then transported African slaves to the Americas ("Middle Passage"), and finally delivered sugar and its derivatives, such as molasses and rum, back to Europe ("Final Passage"). Yellow fever was first recorded in the Americas in 1647 on the island of Barbados and, from that toehold in the Lesser Antilles, the virus spread to tropical lands throughout Central and South America as trade in sugar and slaves expanded. The first, island, occurrence of the fever was followed in 1648 by an epidemic on the Yucatan peninsula of Mexico.

Yellow fever was not restricted to faraway lands, nor jungles, steamy or otherwise. As the North American colonies grew and achieved independence, trade in humans and products brought the virus to the young United States, where 18th

The Silken Thread. Robert N. Wiedenmann and J. Ray Fisher, Oxford University Press. © Oxford University Press 2021.
DOI: 10.1093/oso/9780197555583.003.0010

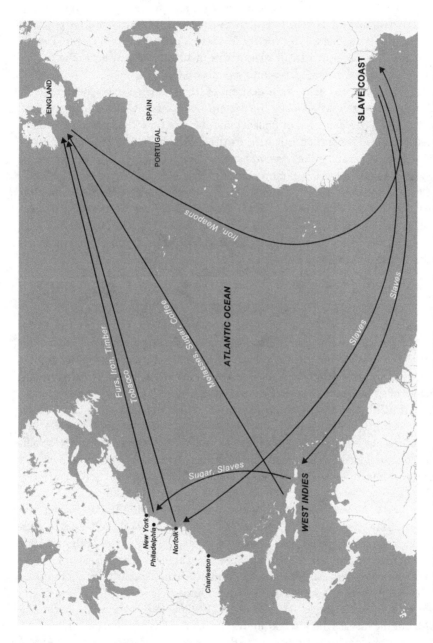

Figure 10.1 Major trans-Atlantic trade routes in the mid-1700s, forming the three legs of the First, Middle, and Final Passages.

Source: Sarah Silva, MapShop.com

Century epidemics of yellow fever wracked the bustling cities of New York and Philadelphia, among other northern US cities. In 1791, the first refugees fleeing the ongoing Haitian Revolution (Chapter 14) arrived in Philadelphia aboard the sailing ship *Charming Sally*. The Haitian Revolution was a significant event in the history of the Caribbean and lands both to the north and south of Haiti, with far-reaching and long-lasting consequences for society. The slave uprising in Haiti targeted white slaveowners, with more than 1,000 killed within just the first few weeks. The stream of refugees continued unabated, reaching a total of 3,000 by 1794.

Philadelphia was not alone as a destination for Haitian slaveowner refugees; ships landed at port cities all along the eastern seaboard as well as along the Gulf of Mexico. The 1793 epidemic in Philadelphia—at the time the nation's capital— sickened countless thousands, resulting in the deaths of nearly 10% of the city's 50,000 residents. Escaping the fever was advised by prominent physician Benjamin Rush, who pleaded: "all that can move, quit the city." His plea led to the hasty exo- dus of the government of the young United States.[1] President George Washington left at the insistence of his wife, Martha. Treasury Secretary Alexander Hamilton fled, but not quickly enough—he was infected before leaving. Although he survived the infection, during his journey he was treated no differently than other sickened refugees—as an outcast.

OUTBREAKS IN THE SOUTHERN UNITED STATES

The 1793 yellow fever epidemic, and its deaths, ended late in autumn, about the time of the first frost, but the disease wasn't gone. The virus persisted in tropical countries and yellow fever continued to arrive in the United States, largely through shipments of slaves from West Africa via the Caribbean. Major outbreaks occurred in the early 19th Century in southeastern US cities: the first record from New Orleans was in 1828 (Figure 10.2). The 1820 epidemic in Savannah, Georgia, killed nearly 700 people, and a repeat epidemic in 1854 killed 650. Farther north, Norfolk, Virginia, was besieged by the disease in 1855 after the arrival in a nearby port of a ship carrying infected passengers; from there, the infection spread and killed more than 3,000.

A significant role in the disease's history was played by the trade of sugar and other goods from the Caribbean along established routes to southern US ports, such as Savannah, Charleston, and New Orleans. Merchant ships of the 18th and 19th Centuries carried more than Caribbean cargo. The ships also carried sailors, whose arrival in southern ports marked the starting point for the establishment of the virus and subsequent onset and spread of yellow fever outbreaks.

New Orleans, for example, had outbreaks of yellow fever every year, beginning in 1828, but the disease did not stay in the city. Outbreaks of the fever, or even the prospect of an epidemic, caused citizens to flee as "medical refugees," trying to escape infection. Many headed north on the Mississippi River. As the 1828 outbreak began, New Orleans residents moved upriver to Memphis, Tennessee, which resulted in a local epidemic after infected travelers landed there. Even though the 1828 Memphis

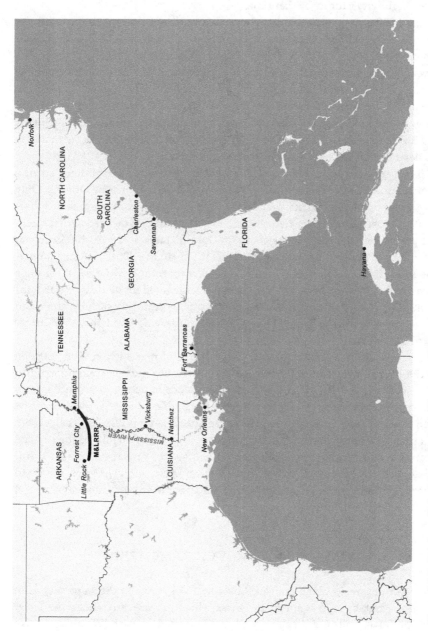

Figure 10.2 Southern United States and the Mississippi Valley, showing proximity to Havana, Cuba. The bold line represents the short route of the Memphis & Little Rock Railroad.

Source: Sarah Silva, MapShop.com

epidemic killed *only* 150 people, that number was relatively large, as there were 650 cases altogether and the population of Memphis at the time was fewer than 1,000. By 1830, the population of Memphis had slowly begun to increase, but only 663 people believed in the city's future by that time.

In the 1800s, some believed that the virus and disease moved to the southern port cities from northern areas along with the flow of immigrants seeking jobs, leading to the fever being called "Stranger's Disease." It is far more likely that the disease was repeatedly reintroduced through Caribbean trade. The periodic arrival of infected sailors and other passengers, plus long, hot, humid summers, allowed the yellow fever virus to maintain itself. Outbreaks occurred along the Gulf Coast nearly every year, reaching epidemic levels in many of those years. In just the three-year period 1853–1855, the disease killed more than 13,000 in New Orleans, with nearly 8,000 deaths in 1853 alone, the worst year to that point.

Cuba experienced yellow fever continuously after its establishment there, and the port of Havana was the origin of many ships that departed for New Orleans. Only during the years of the US Civil War (1861–1865) was New Orleans free of yellow fever. Why those years? First, in April 1861 President Lincoln ordered a total blockade of ports along the coast of the southern United States. In 1862, the occupying Union military, under the command of General Benjamin F. Butler, placed a quarantine inspection station some 70 miles downstream from New Orleans. All ships headed for the city were inspected and their crews detained for 40 days to watch for possible infection. Both the blockade and strict quarantine successfully prevented sickened travelers from arriving in New Orleans and serving as the inoculum for re-entry of the disease.

Prevention of diseases by quarantine was first known as such in the 1300s. To keep the plague from entering Mediterranean port cities, ships were required to anchor offshore for forty days before offloading goods or people, a practice known in Venice as *quarantina giorni* ("space for forty days").[2] In Louisiana, the quarantine station south of New Orleans remained in use after the Civil War, but the practice stifled trade at a time when importation of goods was needed to rebuild the city and its economy. As a result, there were strong incentives to "run" the blockade, meaning that goods and people, mostly from Caribbean ports, still arrived on the sly in southern Louisiana. Quarantine works only if it is enforced, and those arriving "on the sly" rendered it ineffective.

"YELLOW JACK" MOVES TO MEMPHIS

Yellow fever outbreaks along the Gulf Coast in 1878 began in the same way as in other years, but by mid-summer, the disease had spread throughout the lower Mississippi River valley. The epidemic of 1878 affected at least 132 towns in five states: Kentucky, Tennessee, Alabama, Mississippi, and Louisiana. Some sources state that more than 120,000 cases were reported along the river, with total deaths close to 20,000. Others cite 74,000 cases and 16,000 deaths. Regardless, this was a true epidemic throughout the region.

Officials in New Orleans tried a number of tactics to stem the epidemic, such as cleaning the air by burning tar and firing cannons—neither of which succeeded. As many as half of New Orleans residents fled inland to escape the disease, and many of them followed the Mississippi River. Importantly, moving along that watery highway involved unloading and picking up more epidemic refugees at stops along the way.

Unknowingly, some of those refugees also took the disease with them, and each stop became a new hot spot for infection. The prospect of disease cleared out many residents of Vicksburg, Mississippi, and nearby communities, but still claimed more than 3,000 victims. Local quarantines enacted to keep the disease away also prevented trade, and the hit to the local economy in small Mississippi towns meant many of those towns never recovered.

The most significant stop on this inland flight was Memphis, Tennessee, in 1878. Recall that only 663 people remained in Memphis by 1830, two years after the 1828 epidemic. In the next 50 years, the bustling river port grew enormously, and the population swelled to 47,000 (Figure 10.3) despite three successively larger epidemics (1855, 1867, and 1873), which collectively killed more than one-third of the 8,700 that contracted the disease. This time, in 1878, when "yellow jack" arrived and the disease took hold, the city's death toll exploded.

No epidemic before 2020 in the United States can compare to the devastation wrought by the yellow fever epidemic of 1878. News reached Memphis in July that an epidemic had started in New Orleans. The prospect of the disease reaching Memphis and causing another epidemic resulted in the exodus of more than 25,000 of the city's inhabitants. Those who could afford to leave did, and most of those 25,000 who

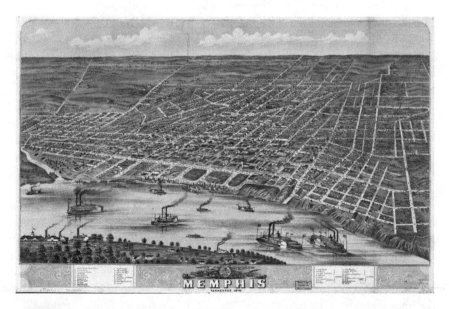

Figure 10.3 Bird's eye view of the city of Memphis, Tennessee, 1870.
Source: Courtesy the Library of Congress

fled did so in an astonishingly short five days. Many escaped in time to avoid the virus; some did not.

Despite checkpoints and quarantines, the virus arrived in riverboat crew and, when the epidemic hit, it hit hard. Of the 20,000 Memphis residents who remained, more than 17,000 contracted the disease and more than 5,000 died—one-quarter of the remaining residents. The number of deaths was one-quarter of all the deaths in the five mid-south states during that epidemic. Put that number 5,000 into perspective: more deaths were recorded in Memphis from yellow fever in a few months than the combined losses of life in the Great Chicago Fire of 1871, the 1905 San Francisco earthquake, and even the 2,200 fatalities from the Johnstown (Pennsylvania) Flood of 1889. In fairness, those three disasters incurred fatalities in a matter of a few days each, not the months it took for the deaths to accumulate in the Memphis yellow fever epidemic.

Life in the hell that was Memphis during the epidemic cannot be imagined today.[3] Pleas from city leaders to President Rutherford B. Hayes were met with indifference or even indignation. Hayes regarded the descriptions of conditions in one entreaty for government help as "greatly exaggerated" and dismissed the pleas.

Food and fuel were hard to find as the city's streets were gradually deserted. The ranks of the police were thinned by sickness: only five of the city's force of 48 officers were well enough to work, ten having died and 33 sickened but surviving. There was little need for police, though, as nearly the only people in the streets were the collectors of the dead. Quick burials followed for the victims, because people thought that corpses spread the disease through "bad air" (Figure 10.4). At the peak of the epidemic, 200 Memphians died each day, causing a shortage of coffins, as many woodworkers had fled the city.

The world of the fever was a world turned upside down. Even with the high summer temperatures, people boarded their windows and burned fires in stoves and fireplaces to keep out that "bad air." The clothes and beds of the dead were taken to the streets and burned. Closed schools were used as hospitals. Houses of victims were disinfected through the burning of sulfur or pyrethrum, but then had to be aired out, as both disinfectants were toxic. Apparently, no one considered that the "airing out" might have provided an opportunity for the disease to re-enter.

Despite a lack of federal government support from President Hayes, aid arrived in different forms. Northern states sent coffins, medical supplies, and tens of thousands of dollars in aid. The Howard Association, formed during the 1855 yellow fever epidemic in Norfolk, Virginia, organized doctors and nurses, who arrived from around the country. Those who came to help included clergymen and other heroic volunteers, many of whom died caring for the ill. Local residents who stayed also offered aid to the sick, including prostitutes who served as volunteers caring for those stricken by the fever. Many died because of their heroic efforts. By the end of the 1878 epidemic—again coinciding with the first killing frost—the city had lost 30,000 residents, either through death or migration, leaving the city in financial ruin.

VICTIMS OF THE FEVER AWAITING BURIAL AT ELMWOOD CEMETERY, MEMPHIS.

Figure 10.4 Burying yellow fever victims during the 1878 Memphis yellow fever epidemic; Elmwood Cemetery.
Source: Frank Leslie's Illustrated Newspaper (1878). Illustration provided by the Memphis and Shelby County Room, Memphis Public Library & Information Center

MEANS OF ESCAPE

The 25,000 people trying to escape the epidemic in Memphis took any mode of transport possible, including riverboats and railroads. This desperate flight incited responses from neighboring states trying equally desperately to keep the disease out. In St. Louis, Missouri, more than 600 kilometers upriver, arriving passengers who were suspected of being infected were conveyed by a specially contracted steamboat into quarantine away from the city. The Governor of Arkansas issued an edict requiring the state militia to guard all entry points into the state, including patrolling the Mississippi River banks near possible crossing points from Memphis.

Despite the militia and quarantine, yellow fever made its way to several small towns on the Arkansas side of the river. The cause of the disease was not yet known, and trade of goods was blamed, especially cotton. Cotton shipments into Arkansas were banned until after the disease abated that year.

In August, Arkansas officials ordered the Memphis & Little Rock Railroad (M&LRR) to stop transporting passengers from Memphis, but the order was ignored and enforcement spotty. On the morning of August 15, the M&LRR train bound for Little Rock was stopped at a station some 10 miles northeast of the town and the complement of more than 130 passengers was removed from the train and held in quarantine for 21 days.[3] Twenty-one days! Other M&LRR trains were stopped at Forrest City, about 90 miles east of Little Rock, and passengers were required to dis-embark and submit to an inspection before being allowed to re-board and continue the train trip. It is no surprise that railroad traffic— both passenger and freight— was severely curtailed in Arkansas because of the bans and inspections.

Pause for a moment and consider the impact of this railroad interdiction in Arkansas, putting it into a historical context. The year was 1878—only two years after the country's centennial and barely a dozen years after the end of the Civil War. The population in the 38 states of the United States was nearing 50 million. Railroads were huge in the United States, less than a decade after the "Golden Spike" hammered into the tracks in Utah tied together the country from east to west.

By 1878, there were more than 81,000 miles of operating railroads in the country. Of that total, only 800 miles of tracks were found in Arkansas and 133 of those miles on the route from Memphis to Little Rock. By 1880 (most rail data were collected every decade in the years in the US census; some data were collected only every 20 years), there were 17,800 freight locomotives, 2,000 passenger locomotives, and more than 11,600 passenger cars. The total national miles of railroads doubled between 1878 and 1890, with 163,597 miles in the latter year. In 1890, more than 492 million passengers traveled an average of 24 miles, for a total of 11.8 billion passenger miles—large figures for a population of 50 million people. The railroads were a major means of moving people, and the railroads moved people away from centers of the 1878 epidemic. So, imagine the disruption of being detained for 21 days or having to de-train, be physically inspected, then re-board to continue the trip.

The little M&LRR that plied the 133 miles from Memphis to Little Rock had its own history. The first tracks were laid in 1854, and the railroad had only 38 miles of track west of Memphis by the onset of the Civil War in 1861. On its first run, the railroad carried 6 bales of cotton. In 1861, the railroad began putting down tracks from Little Rock to the east, ending at DeValls Bluff on the White River. The middle third, 45 miles of swampy lowland between DeValls Bluff and Madison on the St. Francis River, was a challenge to traverse by any means—this was the Big Woods, at one time the largest expanse of forested wetland in North America. The dense swamp flooded periodically and was home to alligators and clouds of mosquitos. It also was likely the last stronghold of the ivory-billed woodpecker, whose double-note call is now a distant memory.

The task of finishing the middle third of the M&LRR route after the end of the Civil War was given to Nathan Bedford Forrest. Anyone with an interest in the US Civil War knows of Confederate General Nathan Bedford Forrest.[4] His lightning-fast attacks are legendary and even were the inspiration for the German military's "blitz-krieg" attacks of World War II. He was also the namesake of Forrest City, site of the 1878 railroad inspections. Completing the railroad through the swampy lowlands of eastern Arkansas could not have happened without a decisive leader and Forrest, with his reputation as battlefield commander, was just that. Whether he used "lightning-fast" campaigns to complete the railroad was not recorded, but the short time required to finish that middle third of the route suggests he brought a similar zeal and leadership to it. Regardless, completion of the railroad in 1871 meant easier movement westward and significantly shortened the 3-day trip from Memphis to Little Rock.

The M&LRR was the railroad whose passengers were quarantined for 21 days just outside of Little Rock and the same railroad whose passengers had to de-train in Forrest City for inspection. How effective were those methods? In a word (actually, two), *not very*. To know why not, we need to understand a little about the disease, the virus that causes it, and how it is spread—and not spread.

CAUSE OF THE DISEASE

Yellow fever virus is classified as a flavivirus, a group of viruses that mostly are vectored by insects and other arthropods. In technical terms, it is a single-stranded RNA virus. The flaviviruses include those viruses that cause some scary-sounding diseases, such as breakbone fever (aka, dengue), Zika, or the nearly unpronounceable chikungunya, and viruses named after places, such as West Nile virus and Japanese encephalitis. Yellow fever virus gave the flaviviruses their name—"*flavus*" is Latin for "yellow" and victims of the yellow fever flavivirus have a yellowed, jaundiced appearance.

Flavivirus evolution is a fascinating topic, albeit one that is outside the scope of this book. Our intent here is only to flavor flaviviruses. Most species (50–70) are organized in a single genus—*Flavivirus*—making it one of the largest virus genera known. *Flavivirus* has surprising diversity too, including viruses that only affect insects, viruses that affect birds or mammals that have no known vectors, and arboviruses that use either mosquitos or ticks as vectors. It seems possible that those flaviviruses affecting vertebrates evolved from the insect pathogens. Interestingly, species using certain hosts are most closely related to each other. Among those viruses using mosquitos as vectors, there are only two groups of mosquitos primarily utilized—*Culex* and *Aedes*—and, moreover, those viruses using *Culex* evolved from ancestors that used *Aedes* about 5,000 years ago in Africa. Important for our purposes, yellow fever virus is just one of several closely related *Aedes*-vectored *Flavivirus*.

The agent causing yellow fever was initially considered a bacterium and inoculations based on that belief were developed, most of which proved deadly to the patient and conferred no immunity. It was not until 1900 that experiments conducted by James Carroll, by then a member of the Havana Yellow Fever Commission, demonstrated that the disease could be transmitted by human serum passed through a filter. The fine-mesh filter would have captured any bacteria and, thus, the idea of a bacterium as the causative agent was dropped, though not without a fight from supporters of the theory. The term "virus" was not yet in use, but Carroll's experiments suggested the causative agent of the disease was smaller than bacteria, and his findings are considered the discovery of the first human virus.

Yellow fever virus causes the disease, and yellow fever can be a nasty disease. At one extreme, its symptoms are mild, even absent in some victims, and surviving victims are immune to subsequent infections. The other extreme includes symptoms that no one would wish on their worst enemy: headache, muscle pain, high fever, bleeding from the eyes, seizures, and delirium. The disease can linger for days, keeping the patient in pain, or it can strike down within two days someone who appears healthy. Some reports describe victims running, screaming deliriously, through the streets shortly before dying.

Yellow fever virus attacks internal organs, especially the liver, which is the cause of the yellow appearance of the patient's skin and eyes. Progression of the disease leads to bleeding, especially through hemorrhaging. Internal bleeding and the accumulation of blood in the stomach is what causes the most striking indicator of the disease's severity: "black vomit." Once contracted, the disease is often fatal; in some situations up to 50–75% of adults infected die. Once the disease is contracted, there is no cure—only palliative care to ease symptoms.

Although the fever was given the colloquial name "Black Jack" because of the "black vomit," another name for the disease had nothing to do with the disease's symptoms. The name "Yellow Jack," in its association with yellow fever, does not refer to the yellowed appearance of victims. Instead, this descriptive name came from the yellow flag—a "Yellow Jack"—flown from the masts of sailing ships to indicate sickness or quarantine.

One longstanding mystery about yellow fever was the fact that, even in the midst of a staggering epidemic, not everyone contracted the disease. Nor was yellow fever fatal for every person who contracted the disease. The disease affected people of every age, sex, race, and ethnicity, but it did not do so equally. In an extensive analysis of the 1854 yellow fever epidemic in the city of Savannah, Georgia, University of Warwick (UK) historian Tim Lockley found differential mortality among races, sexes, ages, and ethnic background. Children were affected less severely than adults; many showed only mild symptoms and many survived. White children made up just under a quarter of the population but represented only 7% of deaths. Men, especially young men, had greater infection rates than women. Male immigrants were more likely to die than native-born Georgian men.

Infection rates among blacks and whites were similar, but blacks were less likely to die than whites. Of 650 deaths from the Savannah epidemic, 14 were of blacks, despite them making up nearly half of the Savannah population. That differential was even more extreme given that a large fraction of whites left the city to escape the disease, and many of them left behind property cared for by black servants.

WEAPONIZED YELLOW FEVER

The horror that was yellow fever—percent infected, percent of infected victims dying, the rapidity of disease progression, and its late-stage symptoms—was known, even if the causal agent was not. At a time when the clothes of infected patients were still considered the means of spread of the disease, yellow fever was eyed as a weapon to be used in war—the ultimate in biological warfare.

Imagine inflicting disease and the terror associated with it against distant targets, whether soldiers, ordinary citizens, or prominent figures. And if you were a Southerner, there was no better target than the figure representing the North—President Abraham Lincoln. So it was that a plot was hatched, to collect soiled clothes from victims of the disease and ship them to northern cities, even to the nation's capital city, Washington, DC. The entire sordid story[5] is one of twists and turns, shadowy figures and international intrigue.

Ironically, the instigator of this plot was a doctor, one who had treated countless yellow fever patients but who also betrayed his Hippocratic Oath. Before the Civil War, Dr. Luke Pryor Blackburn had been an upright citizen in his adopted home of Natchez, Mississippi. There, he was referred to as "The Good Samaritan" for his treatment of those sickened by diseases, especially the poor who could not afford to pay for his medical services. His position as director of the Natchez hospital, and later Natchez health officer, gave Blackburn a social status that provided him the opportunity to mingle with other elite figures, even Jefferson Davis, later to become President of the Confederate States of America. Unfortunately, Blackburn's beloved wife was sickened and died in 1856 from an unknown fever, throwing him into deep depression.

When war broke out, Blackburn was a most-fervent supporter of the South and hated all things Northern, writing in a letter to his brother that he considered all Union soldiers to be traitors. It is not known just when his extreme beliefs manifested in the plot to use yellow fever as a weapon. The details of the depraved plot nearly defy belief and, indeed, the shady characters with their often-conflicting stories make it difficult to pinpoint its true origin and extent.

Regardless, Blackburn's hatred of what he termed Northern aggressors led him to collect clothes from victims, soiled by "black vomit" and believed to hold the unseen disease agents, fill trunks with the clothes, and ship them to Northern cities, such as Boston and New York, where they were to be sold at auction. The trunks were packed carefully, with the intent of inflicting the most damage, and were even stored in the summer heat in the hope of making the clothes more infectious. One trunk, delivered to an auction house on Pennsylvania Avenue in Washington, DC, only a stone's-throw from the White House, was considered by Blackburn to be so infected that it could "kill at 60 yards." It is not clear how, if he could collect those soiled clothes without being infected, the occupant of the White House would be infected. Logic seems to be absent from the tale.

The plot to commit murder by the unknown disease agent shows how desperate was the situation for Southerners in 1864 and the depths of Blackburn's hatred, that it should lead to such depravity, but the plot also illustrates how terrifying yellow fever was at that time. Despite the serious intent to inflict disease and terror from yellow fever, the plot killed no one, at least not from soiled clothes. The true causal agent of the disease would not be known for several decades.

The plot was exposed and Blackburn—then in hiding in Canada—was arrested and tried. Despite his acquittal due to questions of jurisdiction, Blackburn earned the title of "Dr. Black Vomit." In a curious case of atonement and political forgiveness that might rival 21st Century US politics, Blackburn was later elected Governor of the Commonwealth of Kentucky. As governor, his fervent actions to reform prisons and improve conditions for prisoners were controversial and earned him scorn that may even have exceeded that associated with his terror plot. How one individual could be revered as "The Good Samaritan" and a prison reformer and reviled as "Dr. Black Vomit" remains one of the many mysteries associated with the killer that was yellow fever.

THE CARRIER—*AEDES AEGYPTI*

Yellow fever virus is transmitted predominantly by only one group of mosquitos—*Aedes*—and the virus can only be transmitted by a mosquito bite. There are almost no exceptions—very rare reports of transmission from mother to an unborn child have been accepted, and the virus may have been picked up, once, from a contaminated needle. Basically, yellow fever is a mosquito-vectored virus.

Yellow fever mosquitos (*Aedes aegypti*) are called "domestic" (not to be confused with true domestication) because of their ability to live successfully in the presence of humans. This species (Figure 10.5) is one of the "tree hole mosquitos," which means the larval stages develop in the small pools of rainwater trapped in tree cavities. That behavior alone predisposes this species to thriving around humans. Yellow fever mosquito larvae need only need small pockets of water (e.g., tires, buckets, gutters)—the sorts of pockets of water provided by humans and their dwellings.

Yellow fever has three different transmission cycles functioning in different habitats: sylvatic, savanna, and urban. The "sylvatic" cycle occurs within tropical forests where primates (largely monkeys) are bitten by a mosquito and the virus is passed on to other primates. In this cycle, humans are bitten only if they are in the forest. The "savanna" cycle occurs when the virus is transferred by a mosquito either from monkeys to humans or directly between humans. In this cycle, humans are usually bitten in areas bordering forests, as opposed to within a forest. In the "urban" cycle, the virus is introduced to an urban setting by a previously infected human, then transmitted between urban mosquitos and humans. Mosquitos acquire the virus by feeding on an infected person during the narrow window of time when that person is infectious to mosquitos. That narrow window is open from a few days before fever begins to 5 days after the fever starts. Think about that. Infected patients cannot directly transmit the virus to another human without a mosquito.

Most mosquitos don't bite humans, but that fact is lost on *Aedes aegypti*. Wait a minute! Most mosquitos do not bite humans? Correct. And you thought all mosquitos

Figure 10.5 Yellow fever mosquito, *Aedes aegypti*; ovipositing eggs, which are laid individually on a vertical surface, not on water.
Source: Courtesy Ben Matthews

singled out you and you alone as a target? Actually, no. Only a small percentage of the 3,500 species will regularly feed on humans, and *Aedes aegypti* is one of them.

Domestic *Aedes aegypti* have been shown to prefer feeding on humans and also to prefer man-made structures that provide opportunities for both resting and seeking hosts. *Aedes aegypti* females are primarily daytime feeders and only occasionally crepuscular (active at twilight). They are attracted to humans—specifically to the chemicals in odors that humans emit. In other words, they smell you. If you are living and breathing, *Aedes aegypti* females will find you. Just by breathing, you emit chemicals such as carbon dioxide and octenol. When you perspire, you produce other attractive chemicals in your sweat, including lactic acid, urea, and ammonia. If you are alive, breathing, and sweating—maybe sweating because you are worrying about attracting mosquitos—you are unknowingly but effectively attracting mosquitos. You are helping them do their job.

   Aedes aegypti can live as long as a month as adults, but they are killed by cold temperatures. The eggs survive cold temperatures and desiccation and are viable for up to a year. Only females bite and feed on blood. They don't need blood meals to live—they feed on plant nectar—but they need host blood to produce and mature eggs. Female *Aedes aegypti* also feed multiple times during the cycle of egg production, which has important implications for transmission of the virus among humans.

   Yellow fever mosquitos were once considered native to West Africa, but recent studies point to an origin in Madagascar and nearby islands, from where they were carried to the shores of East Africa. From there, the mosquito made its way across the continent to West Africa. Because of its close association with humans, the mosquito has been moved worldwide; the globalization of humans is mirrored in the distribution of *Aedes aegypti*. So, in *Aedes aegypti* we have blood-feeding insects that have adopted human habitations for their residence, that are attracted to the odors produced by humans breathing and sweating, that preferentially feed on humans and on multiple humans at that, and that carry a virus that has killed hundreds of thousands of humans.

FROM AFRICA TO THE AMERICAS

So we've determined that yellow fever mosquitos are effective vectors for yellow fever, but that alone isn't enough to explain their impact on history. You see, to have caused the havoc blamed on them, these mosquitos needed to overcome a severe obstacle, one too great for most mosquitos—they needed to survive a journey across the Atlantic, which often took two to three months. Adult *Aedes aegypti* usually only live for two to eight weeks, which is not enough time to survive a voyage. Larval stages require fresh water, which might seem intuitively plausible. However, barrels of fresh drinking water were insufficient for the extended journeys, meaning the barrels were empty on arrival, with no water for larvae to develop. So larval *Aedes aegypti* aren't the sole answer to the riddle either. What is the answer? Eggs. And the reason relates to an aspect of this mosquito's biology that we already mentioned—they develop in tree-holes.

   Mosquito eggs and egg-laying behaviors (oviposition) are more interesting than one might suspect. Some mosquitos lay their eggs in rafts on the surface of ponds (e.g., *Culex*); others may also lay eggs singly on the water's surface, each with an air pocket so that they float (*Anopheles*); and still others glue their eggs in star-shapes to the underside of floating plants (*Mansonia*). Elephant mosquitos (*Toxorhynchites rutilus*) are attractive, huge mosquitos that are not blood feeders, relying instead exclusively on nectar. Larval elephant mosquitos develop in tree holes where they are predators of other mosquitos. Females of this species hover above the water and fire eggs at its surface. These eggs float, relying on the surface tension of the water and are easily drowned if jostled. Furthermore, they quickly desiccate if removed from the water's surface, as would happen if the water dried up. That's a fine balance indeed, but not likely to be found in water barrels on the turbulent journeys across the Atlantic.

Tree holes are important because they catch and hold rainwater, gradually drying through evaporation until the next rain. These sheltered cavities provide the kind of habitat that permits elephant mosquito eggs to float. The semi-contained cavity of a tree hole retains humidity for a few days before the water evaporates, as would occur in drier regions. With a return of rains, the cavity would be wet again, refilling by enough to create a small, ephemeral pool of water. In the eastern United States, where elephant mosquitos are found, water in a tree hole is unlikely to dry quickly. The same cannot be said for the habitat of yellow fever mosquitos.

Yellow fever mosquitos are tree-hole mosquitos originally from the islands of east Africa. To say the least, their native habitat is dry, punctuated by occasional rains. The cavities of tree holes there would fill when infrequent rains occurred, then dry between those infrequent rains. Yellow fever mosquitos lay eggs singly in a tree hole above the water's surface, not on the water itself (Figure 10.5). Female mosquitos are stimulated to oviposit on the cavity wall when one of their feet contacts fresh water. Their egg requires humidity for only the first few days until the outer layer (chorion) hardens into a protective shell. The embryo develops into the first larval stage, but does not emerge from the egg, and instead rests in that shell, protected from extended dry conditions. The egg can survive desiccation for up to a year, until it is reintroduced to water, when the larva emerges from the wetted egg. In a tree cavity, the return of rains also means the return of favorable conditions, namely water packed with microbes that the larva needs for food.

To yellow fever mosquitos, a barrel with fresh drinking water would have been a perfect analog for a tree hole. A barrel would have provided a semi-enclosed, humid cavity, with wet sides, and as water was used, the barrel would have drained and become dry. The inside wall of water barrels would have been black with mosquito eggs. The adaptation that allowed *Aedes aegypti* eggs to survive periods of desiccation meant the eggs were still viable after the slave ships arrived in the Americas. As barrels were refilled with fresh water, the mosquito eggs hatched, and the recharged water provided nutrients for their development, which required just a few days. Unlike the terrible living conditions provided for the slaves on board, a barrel of fresh water was first-class accommodation for a tree-cavity mosquito.

What were the odds of yellow fever becoming established in the Americas? Very low. An unlikely string of events needed to line up perfectly. If the thread was broken anywhere along its length, the Americas may never have seen a yellow fever epidemic. Let's begin with the end point—the first epidemic in the Yucatan in 1648—and then work backward. To cause this epidemic, a sufficient number of people needed to be fed upon by infected mosquitos, which required a large number of *A. aegypti* eggs surviving a voyage from Africa. The voyage must have had water barrels with enough space above the water level for female yellow fever mosquitos to lay eggs. During the voyage, adult mosquitos would have had to keep the virus circulating by infecting more slaves. At the start of the voyage, the virus needed to be brought aboard, either by adult mosquitos that were carrying the virus or slaves who were already infected. But also remember there would not have been slave ships bringing slaves from Africa to the Americas at all if there had not been sugarcane. The thread that ended in an epidemic of yellow fever in the Yucatan began with sugarcane and its movement on the Silk Roads through Asia to the Mediterranean, and then on to the Atlantic Islands. Remember, too, that without the Silk Roads, sugarcane production technology may not have escaped India. This thread began with silkworms.

If the thread that ran from Asian sugarcane to a yellow fever epidemic had been broken at any point along its length, there would have been no yellow fever epidemic in 1648. Any one rare event is unlikely to occur, but rare events *do* occur. The reason they are rare is because everything has to go right. There were an estimated 36,000 slave shipments from Africa to the Americas, providing more than enough opportunities for a rare event to occur. And occur it did. Sugarcane, slave ships, and the Silk Roads, converged, and brought yellow fever and yellow fever mosquitos to the Americas.

THE STATE OF KNOWLEDGE

We know all that now. It was not known in the 1800s, at the time of the epidemics. People thought the disease was carried in shipped goods, hence, the ban on shipping cotton and the quarantine of other goods. And, so, the inspectors at Forrest City—what did they look for? Inspectors could not detect the disease in people who were not showing symptoms. Because no one knew about mosquito bites, inspectors also could not know about a pre-fever period when the victims were infectious. Inspection would clearly detect anyone in the throes of yellow fever, but those people were likely too sick to make the train journey. However, a lack of symptoms was not proof that an individual was not infected. Absence of evidence is not evidence of absence. Some victims of the disease could appear healthy but be dead 48 hours later. Plus, those still alive a week after the onset of fever were no longer infectious.

Banning the shipment of cotton or other goods was not only ineffective, it harmed local economies that were already reeling from the loss of laborers. Worse, it focused on the wrong culprit, lulling people into complacency. Inspections were of little value; freight bans of no value; the 21-day quarantine was disruptive but effective, though no one knew why. At the other extreme—far from complacency—the human reaction to the occurrence of an outbreak could cause near-panic. In 1879, with the previous summer's epidemic still fresh in the collective memory, there was an outbreak in the same Forrest City, Arkansas. When word got out about the outbreak, the population of Forrest City dropped from 1,300 to 70, as people fled before they, too, added to the death toll.

Figure 10.6 Dr. Carlos Finlay, Cuban physician; his experiments were instrumental in understanding the transmission of yellow fever.
Source: Wellcome Images/CC BY 4.0

Mosquitos, not "bad air," were proposed as the vectors of yellow fever by the Cuban physician and epidemiologist, Dr. Carlos Finlay (Figure 10.6). His announcement in 1881 was met with ridicule, if it was even acknowledged at all. It wasn't until nearly 20 years later that Finlay's hypothesis was supported by the US Army Medical Commission, among its members Major Walter Reed (Figure 10.7), earlier one of the vocal skeptics of Finlay's idea.

Why were mosquitos not considered earlier as the culprits? In the southern United States, mosquitos in the often-swampy towns and countryside were just part of life. In that regard, mosquitos were no different from the towering cypress trees, meandering bayous, wetlands, or the sky itself. Mosquitos just *were*. Many of those low wetlands, especially where humans lived, held standing water and were fetid and foul-smelling. Wetlands produced "bad air." The name for another mosquito-borne disease—malaria—means "bad air." Bad air as a cause of yellow fever just made sense. But it was just wrong. Finlay's studies showed the real cause, later supported by collaborations with Major Reed. Recognition of the importance of the finding was not shared equally. Ever read about Carlos Finlay? Ever hear the name? How about Walter Reed?

Figure 10.7 Major Walter Reed.
Source: OHA 83 Woodward Photograph and Photomicrograph Collection, WW 2545b. Otis Historical Archives, National Museum of Health and Medicine.

Consider this: would you have suspected mosquitos as vectors? What made Finlay think about mosquitos in the first place? Everyone knew the fever was caused by "bad air" and that it went away when colder weather arrived. Cold weather and autumn frosts were not likely in Carlos Finlay's Cuba. Maybe he saw the world differently, but he certainly didn't simply accept the status quo. Instead, he brought careful observations and simple experiments carried out over several decades to the problem. His experiments included inoculations of human volunteers, trying to demonstrate that he could produce cases of yellow fever and that the inoculations could induce immunity. Criticism of his experiments included a lack of a control to exclude other sources of the diseases and that the fever failed to develop in some of those inoculated.

Finlay's ideas were not accepted for more than 20 years but, once later experiments uncovered the role of *Aedes aegypti* as a vector of yellow fever, suddenly the cycles of disease and its spread made sense. Recall, the epidemics in the United States abated in the late autumn months. In extreme epidemics, the mosquito vectors simply were not finding enough new, uninfected human hosts to spread the virus. And in the more-temperate areas of the United States, the epidemics blinked out after the first good frost, a frost that would have killed the adult *Aedes aegypti*, eliminating that season's vectors. Though, the mosquito population wasn't gone—the eggs over-wintered and hatched the next year.

That knowledge was not available in the late 1800s. Once the disease was known to be transmitted by mosquitos, how did people respond? How would you have responded? Recognizing the role of *Aedes aegypti* as vector of yellow fever led to major changes in lifestyle and modification of human surroundings. Various methods were used to eliminate the breeding grounds of the mosquitos, particularly in and around human habitation. These included draining wetlands, covering water barrels, and eliminating open sewers to reduce mosquito habitat and eliminate suitable places for egg laying, thus breaking the mosquito's life cycle. These effective tactics still form the solid basis of mosquito vector control today.

UNDERSTANDING

Mortality patterns uncovered in Lockley's study of the 1854 Savannah epidemic make sense in light of mosquito biology. His analysis showed there was greater mortality among males than females, and he concluded that there were two possible explanations. First, women left Savannah in greater numbers than men, as many husbands sent families away for safety while they remained behind. The second explanation he offered was due to the mosquito's attraction to hosts and opportunities for bites. Men in Savannah were more likely to be working outdoors, sweating under the hot Georgia sun, maybe even working shirtless.

Aedes aegypti is attracted to the odors in human perspiration, is active during the day, and bites exposed skin—all present in the mid-century working conditions in Georgia. Many of the working men were slaves, but others were young immigrant males, whose only job option was physical labor. A large number of Savannah's immigrants were from the cool climates found in Ireland—for them, the hot summer sun would have been almost unbearable and would have led to them working shirtless.

The immigrant men were poor, mostly living in the poorer neighborhoods, usually near standing water and with poor sanitary conditions. Young male immigrants made up the largest group of victims—nearly two-thirds of the fatalities. Women, in contrast, were more likely to wear long sleeves and long skirts, no doubt uncomfortable, and remain indoors when possible to avoid the summer sun. All those factors reduced the opportunity for *Aedes aegypti* to bite and infect the women of Savannah. Mosquito biology, coupled with human behavior and social conditions, helps to unravel the tangle of yellow fever disease patterns.

The realization that *Aedes aegypti* served as the vector of yellow fever also led to the development of a vaccine that was effective for preventing the disease. In 1937, Harvard virologist, Max Theiler developed the first yellow fever vaccine, which was an enormous leap in the war against the disease. That tactic—vaccination—is still effective against yellow fever today. Theiler, who was awarded the Nobel Prize in Physiology or Medicine in 1951, was the only person associated with the study of yellow fever to be so honored. In addition, Theiler was South African and the first African-born Nobel laureate, although his ancestry was Northern European.

Ironically, the Nobel was awarded to an African-born scientist working in America on a devastating disease carried from Africa by other African-born people, the slaves that were transported to America.

Why African slaves? There were far more reasons than can be covered here, but one prominent reason was yellow fever, vectored by the yellow fever mosquito, *Aedes aegypti*. Although African slaves were susceptible to the virus and its manifestation as disease, fewer died than indigenous people who were enslaved, and so were valued as slaves to work where yellow fever was present. African slaves also were less likely to die from yellow fever than European immigrants or their American descendants. That differential mortality affected US history in the 1800s. Kathryn Olivarius, Stanford professor of history, has pointed out the far-reaching social implications of yellow fever.[6] The disease did not just kill, but it led to creation of a social structure, or hierarchy, based on whether one had been exposed to the disease and survived.

Recall, children suffered lower mortality to the disease than did adults. Children growing up in their native southern cities were exposed to yellow fever. Those who survived were immune to the disease, thus were considered "acclimated"—we would now use the term resistant—and, as adults, could ascend the social hierarchy. Immigrants were susceptible to the disease, because they had not grown up with the virus and were not acclimated. Some of the native, acclimated children became the next generation of leading citizens, their prominence enhanced by their being acclimated. Hire them; loan money to them; marry them—the disease wasn't going to kill them.

For whatever reason, black slaves from West Africa also were less likely to die from the disease, and so were acclimated. But acclimation did not grant them upward passage in the social hierarchy. Instead, being "acclimated" to the disease made them more valuable as slaves, as evidenced by the higher price paid for them. Ironically, the reduced likelihood of death from yellow fever did not benefit West Africans. Instead, survival of the disease helped maintain the focus on West Africa for the slave trade, the trade that was maintained at least in part by the role of yellow fever and, in a best-supporting-actor role, mosquitos. That supporting role would not be known until the 1900s but, silently, unrecognized, mosquitos fanned the flames of yellow fever epidemics and, by extension, the African slave trade.

NOTES

1. Seen on the internet: Question: "What was the worst yellow fever epidemic in the U.S.?" Answer given by a reader: "1793 in Philadelphia." Follow-up question from a reader: "Who was president during that epidemic?" Answer given by a reader: "Franklin Delano Roosevelt." See: https://www.answers.com/Q/When_was_the_yellow_fever_epidemic, accessed March 4, 2020.

   Although the answer "1793 in Philadelphia" could be contested, depending how one defines "worst," the answer that Roosevelt was president could be correct only if Roosevelt was president for 152 years. He was definitely the longest-serving president, but 152 years is a little much. This points out the veracity of the adage, "Don't believe everything you read on the internet," an adage we have tried to keep in mind while writing this book.

2. In this usage, the word quarantine, dating to the 1660s, has been modified multiple times over the course of its etymological history, see: www.etymonline.com/word/quarantine. The word is derived from the Italian *"quarantina giorni,"* meaning "space for forty days"; *quarantina* comes from the Italian *"quaranta,"* meaning "forty," which itself is derived from the Latin *"quadraginta." Quarantina giorni* was the time period for which all ships were held offshore and isolated before the crew and passengers were allowed ashore.

3. A 21-day quarantine could not have been imagined until the pandemic from the arrival in the United States in 2020 of the COVID-19 virus. Many parallels existed between the 1878 epidemic and the 2020 pandemic. However, the original quarantines in 14th-Century Venice were forty days. By comparison, that makes the 21 days' quarantine near Little Rock seem only mildly disruptive.

4. Before the Civil War, Nathan Bedford Forrest had been one of the largest slave traders in the southern United States, selling as many as 1,000 slaves in a year. Forrest was a decisive Confederate general and battlefield strategist, but he was reviled as a slave trader and despised as a leader of the post-War Ku Klux Klan.

5. For the fascinating and detailed story of the first bioterrorist plot to be carried out in America, see "The Gentle, Kindhearted Bioterrorist": Luke Pryor Blackburn and the Yellow Fever Plot (1864–1865)," in J. Michael Martinez, *Terrorist Attacks on American Soil: From the Civil War Era to the Present* (Lanham, MD: Rowman and Littlefield, 2012).

6. From an interview on National Public Radio, October 31, 2018.

CHAPTER 11

ৎ৵ঌ

The Caribbean, Carlos Finlay, Walter Reed, and Serendipity

The history of yellow fever is defined by the history of slavery. And the history of slavery is largely built upon the history of sugar. History can be considered a chain of events, one event leading to the next and then to the next; and many events do connect in a chain like that. But, a better metaphor may be a web, much like a spider web, that underlies the connections in history, the seemingly disparate occurrences or facts that converge, unanticipated. The connections, the webs, are there, once we start to look for them.

In Chapter 9, we followed sugar to Madeira, São Tomé, and Brazil, and the parallel movement of slaves. Sugar and slavery in Brazil set the stage for yellow fever, which first appeared in the Americas in 1647 and 1648. Sugar and slavery were not as closely linked with yellow fever in Brazil as they were in the Caribbean, especially in the 18th and 19th Centuries. Sugar and its influence on yellow fever were also linked closely to construction of the Panama Canal, nearly leading to its failure. Really? Sugar? When—or if—you think of yellow fever, you likely don't think of sugar. Perhaps you remember yellow fever in the historical tales of the Panama Canal, the engineering marvel of the early 20th Century? OK, fair enough. That is the subject of Chapter 12. Or Walter Reed? Yes, surely. But sugar? Yes. Follow the silken threads just a bit. The story will become a little more complex, but keep following the thread. The origins of the successful construction of the Panama Canal go back at least to the arrival of sugar in the Americas, especially its movement to the Caribbean islands.

Sugar dominated development in the Americas, and sugar was that silken thread that formed the web connecting the economic and social history of the Caribbean and the United States. The earlier history of sugar and its arrival in the Americas is detailed in Chapter 9. Sugarcane, most likely Creole cane from Asia, accompanied Columbus on his second voyage to the New World, where it was planted on

The Silken Thread. Robert N. Wiedenmann and J. Ray Fisher, Oxford University Press. © Oxford University Press 2021.
DOI: 10.1093/oso/9780197555583.003.0011

Hispaniola. Columbus was well connected with Portugal and Madeira. He even married the daughter of one of the governors of the Madeiran island Porto Santo, and Columbus and his bride lived on the island for a time. Given the connection between Portugal and Creole cane, and the later cultivation of Creole cane throughout the Caribbean, the assumption of a connection between Columbus and Creole cane is pretty well supported. Reaching that conclusion means that sugarcane was introduced somewhere else in the Americas before being introduced to Brazil. But after its introduction by Columbus, sugarcane was restricted to some Caribbean islands, and large-scale cultivation of sugarcane did not occur until Brazil was colonized. Columbus did not return to the Americas on subsequent voyages because of sugarcane. It was gold that he sought.

Spanish voyagers journeyed across the Atlantic hoping to return laden with the riches of gold, and the conquistadores did indeed plunder and mine gold to bring back to Spain. Easy sources of gold were played out within a few decades after Columbus arrived, but sugar, even with its humble beginnings, eventually replaced gold as a desired and valuable treasure. Sugar was at one time even called White Gold, and in the Caribbean islands of the Americas, gold blinked out but White Gold lived on.

In 1630, the Dutch took control of northern Brazil and claimed "New Holland" as theirs until the Portuguese reclaimed the territory in 1654. Earlier, in 1640, the Dutch had taken sugarcane from Brazil to the British colony of Barbados and established its production there. Before 1640, Caribbean farmers were growing cotton and tobacco, but the large-scale production of those crops in North American colonies drove down prices and cut into profits. Owners of large Caribbean plantations therefore switched to growing sugarcane, and soon nearly every Caribbean island had sugar plantations and mills that refined the cane (Figure 11.1).

In the late 16th and 17th Centuries, the major European empires carved up the Caribbean into their own domains, each with colonies on their own sugar islands. There were some exceptions. The Dutch, early key players in the Caribbean, subsequently sought their riches primarily from shipping (especially slaves) or from colonies elsewhere in the world but retained their mainland colony of Suriname. Those European powers that stayed strengthened their holdings. The French had colonies in Martinique and Saint-Domingue (later named Haiti). Spain had Cuba and Puerto Rico, as well as mainland colonies that blanketed the continents of North and South America. Britain's deep ties to the Caribbean began in 1627 with Barbados as their toehold in the West Indies, soon after adding Jamaica to their island colonies.

By the 1740s, the primary producers of sugar were the island colonies of Jamaica and Saint-Domingue. At the end of that decade sugar was the most valuable commodity traded in Europe, representing 20% of all imports into the continent. In the 25 years between 1766 and 1791, the British West Indies produced and exported more than one million tons of sugar. At one point, 93% of all exports from Barbados was sugar, making Barbados the richest of all the European colonies in the Caribbean.

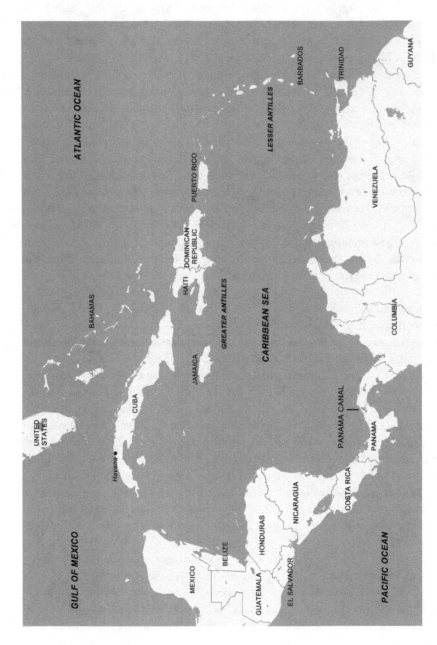

Figure 11.1 The Caribbean Sea, Caribbean Islands, and Central America.
Source: Sarah Silva, MapShop.com

SUGAR CONNECTS TO SLAVERY

Huge profits from sugar plantations ensured sugarcane's dominance as the primary crop produced in the Caribbean during the 18th and 19th Centuries, with cane plantations producing more than 80% of the sugar consumed in Western Europe. And consume it, they did, especially in England. At the beginning of the 18th Century, British people consumed 1.82 kg (4 lbs) of sugar per person per year. At that same time, the British consumed coffee more than tea, the tea that was drunk being mostly green tea.

Within a short time, black tea became more desired and the increase in sugar consumption in Britain went hand in hand with the rising popularity of tea, especially black tea sweetened by sugar. By the end of the 18th Century, the per capita consumption of sugar in England had increased more than fourfold to 8.2 kg (18 lbs) per year (as an aside, Americans in 2019 consumed 33.3 kg (73 lbs) per person per year). Previously, only the wealthy could afford sugar but, as production increased in the British West Indies, lower prices meant the commoners of Britain could buy more sugar.

The demand for sugar in Europe was met by plantations in Caribbean colonies, and large sugar plantations required a large labor force. The word "force" has multiple meanings: in addition to serving as a noun (e.g., a body of people available for a specific purpose), the verb "force" means to compel by physical means. Both meanings applied in this case: the "labor force" for sugar plantations was "forced labor." In the Caribbean, the native Arawaks had been all but eliminated after the arrival of Europeans. The absence of a native people to work on plantations meant other workers were needed, and that need was met by West African slaves.

Movement of slaves to the New World by Portuguese traders was first recorded slightly more than a decade after Columbus's first voyage, but the slave trade really began in earnest when sugar moved to the Caribbean. That was also when the British Royal Africa Company was formed, in 1672, initiating the movement of African slaves to sugar plantations in the British colonies. Also, the origin of the slaves shifted north to countries near the Gulf of Guinea, aligning with British holdings in their colonies along the Slave Coast in West Africa (Figure 9.4).

Of the estimated 10 million Africans taken as slaves to the Americas, as many as 2 million were transported in British ships; roughly 70% of those slaves were destined for sugar plantations. Working on sugar plantations was brutal work, often deadly, requiring more and more slaves to replace those who were killed or maimed in cane production. Even more likely, the slaves died from diseases that were prevalent on the tropical islands. So, shipments of slaves continued unabated from West Africa to the West Indies.

The trade in humans made Britain rich and prosperous, especially the financial sector: the banks and insurance companies that underwrote and insured slave voyages made London a very wealthy city. The slave trade was just one leg of a trade triangle (Figure 10.1). The entire triangular trip made money: (1) movement of goods manufactured after the Industrial Revolution to Africa; (2) movement of humans to the Caribbean and North America in the "Middle Passage"; and (3) movement of sugar, rum, furs, and timber to Britain and Europe, completing the triangle. Huge profits were realized from each leg of the triangular trade, and those profits drove the continued forced movement of enslaved Africans to the Caribbean.

Long before the luster of the British Empire faded, the British Sugar Empire began its own decline, but not due to battles or enemies. Instead, public sentiment about slavery changed. The Slave Trade Act was passed by the British parliament in 1807, but it was not until the Slavery Abolition Act of 1833 that the British got out of the slave trade. Abolition of slavery meant that British colonies had to grow sugar without slave labor, which cut profits and greatly curtailed British sugar production. Consequently, sugar production in plantations in British colonies dropped off but increased at plantations on the Spanish colonies of Puerto Rico and Cuba, where slavery still existed. Cuba became the leading producer of sugar in the world, with Puerto Rico ranked second. In 1860, Cuba alone produced one-third of the world's sugar, which required a large slave population—at that time, slaves in Cuba numbered 400,000.

How important was sugar in history, even just the narrow slice of history that overlaps with entomology? We have presented its role in perpetuating the institution of slavery. Some historians believe that sugar may have been partly to blame for the British losing the Revolutionary War. British owners of Caribbean sugar plantations in the late 1700s feared slave revolts and demanded military protection, even during the war. That protection came at a high cost, however, as it kept British soldiers out of decisive battles against the revolutionaries in North America; resulting in British defeats against the colonial army and success for the 13 American colonies in their fight for independence.

Because of the massive profits that could be made, the trade in sugar and slaves has been considered to have enabled the Industrial Revolution to occur. Manufactured goods, especially weapons, produced in British and other European factories were traded extensively, with some sent to Africa, forming one leg of the Trade Triangle. There, the manufactured goods were exchanged for slaves and the weapons were used to capture slaves. And on the next leg of that trade, the infamous Middle Passage, slave ships unknowingly carried *Aedes aegypti* and the yellow fever virus from West Africa to the New World, as freeloading passengers. The mosquitos, and the fever, passed from island to island and then to Central and South America, as traders moved among colonies. When the British abolished the slave trade, Cuba became the leading destination for slave ships and was the site of continuous outbreaks of yellow fever for the rest of the 19th Century. As related in Chapter 10, the port of Havana was the origin point for many ships sailing to New Orleans, carrying Caribbean goods, humans—and yellow fever.

EARLY PLAYERS

Ironically, although Cuba was the source of the yellow fever epidemics in the United States, Cuba also produced Dr. Carlos Finlay (Figure 10.6), the Cuban physician and epidemiologist whose observations, hypotheses, and experiments showed that mosquitos were the vectors that transmitted the disease. His theories—ignored, forgotten, rediscovered, and then accepted—enabled the combating of the disease and, later, the successful completion of the Panama Canal (Chapter 12). Sugar to the Panama Canal. The silken thread continues to connect.

Walter Reed (Figure 10.7) is often credited with solving the mystery of yellow fever transmission, or at least he is associated with the solution. Reed himself identified the source of the ideas that led to the solution—Carlos Finlay. Juan Carlos Finlay was born in Cuba in 1833 to a Scottish father and French mother, who later changed their names to better fit their Cuban home. The young Finlay spent much of his time as a teen studying in Europe, in both Germany and France. After he returned to Cuba, he—like his parents before him—legally changed his name to further embrace his Cuban identity. Juan Carlos Finlay became Carlos Juan Finlay. Did we miss something? Changing from Juan Carlos to Carlos Juan somehow carried a stronger sense of Cuban identity? OK. Again, maybe we missed something there. Regardless, Carlos Finlay attended medical school in Philadelphia, and after completion of his medical training, he returned to Havana and opened a practice as an ophthalmologist.

Finlay was intrigued by yellow fever and its causes. One might say he was obsessed, though that may be a little harsh. Let's stick with intrigued. Yellow fever ran rampant in Cuba, and Finlay wanted to know why. He sent a paper in 1865 to the Academy of Sciences in Havana expounding the theory that weather played a role in yellow fever disease. Throughout his career, Finlay submitted more than 100 articles either to medical journals or as presentations at conferences: 70 focused on yellow fever. Because of his knowledge of the disease in Cuba and his ideas, Finlay was named in 1879 by Spain to serve on the first American Yellow Fever Commission in Cuba.

One of Finlay's fellow commission members was renowned bacteriologist Dr. George M. Sternberg of the US Army (Figure 11.2). Sternberg's interest in yellow fever was more than academic. While posted in Florida earlier in his career, he was sickened by yellow fever, after which he became obsessed with finding its cause, which he "knew" to be a bacterium—he was, after all, a bacteriologist. When the only tool you have is a hammer . . . . Germ theory was the prevailing explanator for many maladies of that time, but "germs" did not include pathogens smaller than bacteria. Sternberg's expertise was microscopy, particularly photomicroscopy—taking magnified pictures of pathogens using specialized microscopes. His expertise, abilities, and perspective were important to the commission's work.

While in Cuba, Sternberg studied the blood of yellow fever patients through the microscope and produced more than 100 images. But the bacteria responsible for yellow fever eluded him, no matter how many slides of victims' blood he examined. Sternberg's later career was filled with acclaim and success, including finding the bacteria that caused pneumonia and serving as the army's surgeon general. Despite his determined efforts and certainty that yellow fever was caused by bacteria, the reason he failed to find the causative agent was because the disease was not in fact caused by bacteria. His obsession with germs and his focus on bacteria made him miss the clues that were in front of him.

Figure 11.2 Brigadier General George M. Sternberg, US Army Surgeon General.
Source: U.S. Government/Public domain

A SINGULAR OBSESSION

Finlay was similarly obsessed, but with a different focus. He believed that the causative agent was spread through blood, and so he questioned how blood—more specifically, unknown "virulent particles" in blood—could be transferred from one person to another, from a sick person to a healthy one. He was not the first to consider mosquitos as the agents that transferred infectious agents. French scientist Louis-Daniel Beauperthuy suspected that mosquitos were responsible for the spread of yellow fever: he thought they picked up germs from filth. In 1848, American physician Josiah Nott suggested the germs causing yellow fever could pass through the air, like an insect. Although both men were wrong, there was a kernel of truth to their ideas and that truth would be expressed in the ideas of Carlos Finlay.

Finlay combined what he learned from examining Sternberg's microscopic images, his own observations and experiments, and the available epidemiological data to conclude that spread of yellow fever required a combination of three factors or phenomena, all three based on mosquitos. First, a person already infected with yellow fever must be bitten by a mosquito, thus infecting the mosquito with virulent particles. Second, the mosquito must be able to live long enough to bite a new host.

And third, at least some of the uninfected humans subsequently bitten must be susceptible to the disease.

In 1881, Finlay gave presentations outlining his theory at conferences, first, in February in Washington, DC, and in August to the Academy of Sciences of Havana. He posited that a mosquito (at that point named *Culex fasciatus*, later known as *Aedes aegypti*) was the yellow fever vector: mosquitos bit an infected victim with the disease and subsequently bit a healthy person, who became infected. The response of those in attendance was underwhelming. Often the term "hearing crickets chirp" is used to denote a professional presentation that is not well received, with the audience either asleep or biting their collective tongues. In the case of Finlay and his presentations, one might have "heard mosquitos buzz" instead of crickets chirping.

Finlay's mosquito theory was either met with disdain or totally ignored—the medical community already knew that yellow fever was caused by "filth." Germ theory had the medical community firmly in its grip—far better than belief in "miasma" as causal agent—but germs at that time meant bacteria. Even as he spoke, Finlay knew his ideas would be dismissed, saying in his presentation, "nothing less than an absolutely incontrovertible demonstration will be required before the generality of my colleagues accept a theory so entirely at variance with the ideas which have until now prevailed." The evidence he provided must not have risen to the level of "incontrovertible demonstration," for it was not enough to shake the skepticism of the scientific community.

Finlay had in mind that the mosquitos later known as *Aedes aegypti* might be the vector of yellow fever. They fed on the blood of their hosts and found and bit multiple hosts. They also were inactive at lower temperatures, fitting both his observations and the temporal patterns reported of yellow fever outbreaks elsewhere. Finlay continued to observe and conduct experiments for the next 20 years, producing results that supported his ideas. Still, he was ignored. At that point, no one knew about viruses, let alone what could transmitted them. Although viruses had been treated by Edward Jenner in 1796 (smallpox) and Louis Pasteur in 1885 (rabies), the causal agents were not known.

UNKNOWN, BUT CONNECTED, DISCOVERIES

It is remarkable how easy it can be to connect historical events or discoveries after the fact. Equally, it is remarkable how difficult it can be to make those same connections at the time. Transferring knowledge gained from a discovery to another context requires someone putting together the key pieces, which requires being aware that there even are key pieces to fit. Finlay's idea of an insect vectoring the agent causing yellow fever seemed outlandish, but his idea should not have been treated as such.

Finlay's 1881 presentations in the United States and Cuba were not the first suggestion that mosquitos played a role in diseases: there were the early ideas expressed by Beauperthuy and Nott, those these were anecdotal, not experimental evidence.

But the transmission of a disease agent by an insect actually had been demonstrated four years before Finlay gave his presentations. That demonstration occurred half a world away, in China.

Patrick Manson, a young English scientist, showed in 1877 that the mosquito *Culex pipiens* was the vector of a filarial worm that caused Bancroftian filariasis. Clearly, filarial worms were not involved in yellow fever: Sternberg and Finlay would have seen them in their examination of microscope slides. Filarial worms were not the issue. Instead, the issue, the possible connection, was that the worms that caused a disease could be carried by a mosquito. Transmission of yellow fever by mosquitos was, in other words, quite possible.

Finlay probably did not know about the relationship between *Culex pipiens* and filarial worms shown in a study in China. Likewise, while carrying out his limited experiments in Cuba, he was probably also unaware of another discovery made much closer to home, in the United States, that a disease was vectored by an arthropod— albeit a tick, not a mosquito.

Texas cattle fever would not seem to figure into the story of yellow fever—cattle fever didn't affect humans, nor was it carried by mosquitos. But, the fatal disease threatened the US cattle industry in the late 1800s. It was especially prevalent after cattle drives mixed together Texas Longhorn cattle with midwestern cattle in common grazing areas. Ranchers figured prominently in the story in several ways—they noted the presence of ticks on their cattle, their careful observations led to the association of ticks with the fever, and they demanded action by the US government.

The US Department of Agriculture's Bureau of Animal Industry was created in 1884, with veterinary surgeon Daniel Elmer Salmon as its first chief. Salmon believed that the ranchers' observation about ticks merited investigation, despite—yet again—disbelief by mainstream scientists (what is it about mainstream scientists?), and he assigned three young scientists to tackle the theory. Through parallel, connected studies, Theobald Smith, Frederick Kilborne, and Cooper Curtice, uncovered the biology of the cattle tick, *Boophilus microplus*, discovered the protozoa that caused cattle fever, and demonstrated that cattle ticks were necessary for disease transmission. The 1893 report by Smith and Kilborne, stating that the vector of the protozoan was the cattle tick, demonstrated transmission of a disease agent from an infected to an uninfected organism by an arthropod, the cattle tick in this case.

We know all that now. Insects had always been seen as inconsequential, as nuisances, but nothing more. The idea that insects (or other arthropods) could have significant impacts was slow to develop. The discoveries of the mosquito vector of filariasis and the tick vector of Texas cattle fever occurred at roughly the same time as Finlay's experiments. But transmission of ideas in the late 1800s was slower than transmission of diseases. Even if the reports were generally known, the studies were so different that they may not have initiated a connection. Someone at the time still would have needed to make the leap to a mosquito as vector of a disease that had nothing to do with filarial worms or ticks. But that someone was missing.

SHORT, DEADLY WAR

February 1898 marked a time of great change in the history of the Caribbean, the United States, and yellow fever. Tension was in the air between the United States and Spain. On the evening of February 15, the armored cruiser *USS Maine* was anchored quietly at harbor in Havana. At 9:40 p.m., that quiet was broken by a thunderous explosion, resulting in the sinking of the *Maine* and loss of 260 (some sources say 268) sailors, more than two-thirds of the crew. Although blame was never formally placed on Spain, public sentiment clearly supported the claim. Facts be damned, the public wanted revenge! The clamor for war grew. On April 25, 1898, President William McKinley requested that Congress declare war against Spain. Curiously, that declaration came the day *after* Spain had declared war against the United States.

Yellow fever was already part of the story. The explosion on the *Maine* resulted in many injured sailors, burned or with bones broken. Rather than move the victims to local hospitals for treatment, the injured sailors were moved to sick bays on other ships. Even though the care they received was limited and not as thorough as would have been given in Havana hospitals, commanding officers feared sending the injured to hospitals that held many patients afflicted with yellow fever. On shore, yellow fever was playing a devastating role even before the arrival of US troops in Cuba. When war broke out in 1898, the Spanish army of 230,000 troops in Cuba had been reduced to only 55,000 who were healthy enough to fight. In the three years leading up to the war, the fever had already killed 16,000 Spanish troops.

US military leaders knew about yellow fever from the epidemics in the southern United States, and they knew the association between a summer rainy season and an outbreak of yellow fever. Nonetheless, the US Fifth Army invaded Cuba on June 22, 1898, at the onset of the four-month rainy season. The fighting was fierce, if brief— 113 days. Lieutenant Colonel Theodore Roosevelt, who had been assistant secretary of the navy until May 10 of that year, gained fame by serving as commander of the 1st United States Volunteer Cavalry, known as the "Rough Riders" (Figure 11.3). The regiment was cavalry and Rough Riders in name only, as horses were not shipped to Cuba with the unit—other than a horse for Roosevelt. In reality, they amounted to Rough Walkers.

After US troops defeated Spanish troops in bloody battles at Santiago and the San Juan Heights (Kettle Hill and the famous San Juan Hill), the real battle and test of the mettle of the army began—against yellow fever. Roosevelt understood the dangerous conditions for the troops and wrote, in a letter to Secretary of War Russell Alger, that keeping fighting troops in Cuba would lead to a disaster. Military doctors estimated that if left in Cuba through the rainy season, over half the soldiers would die. Alger did not heed Roosevelt's advice and troops began to fall from the fever. Nearly 90% of the 3,000 casualties of the war were from disease, mostly yellow fever. The armistice signed on August 12 ended the short war and the fighting, though not the casualties.

About 50,000 Americans remained after the war as an occupation force, and diseases continued to ravage the troops. Major William Crawford Gorgas, serving as the US Army's chief sanitary officer in Havana, proclaimed the diseases were caused by

Figure 11.3 Theodore Roosevelt and Rough Rider officers during the Spanish-American War. Colonel Roosevelt is in the front row, fifth from left.
Source: The Rough Riders, 1899/Public domain

filth, dirt, and unsanitary conditions. The unsanitary conditions no doubt affected the occurrence of typhoid, but not yellow fever, and the fever returned in July 1899—the start of the wet season. Army doctors were able to treat malaria with quinine, but they had no treatment that was effective against yellow fever.

Many of the new soldiers arriving to fill the ranks of the occupation force's 10 regiments were recruited from Southern states that had already experienced epidemics, in the mistaken belief that because of their origin they would be immune to yellow fever. Many of those soldiers, too, were soon stricken, despite all the army's efforts at sanitation and quarantine. The reoccurrence of the disease and its effects on the remaining military prompted Army Surgeon General George M. Sternberg to appoint a four-man commission, in 1900, officially titled the United States Army Yellow Fever Commission and made up of specialists in infectious diseases, to investigate the cause of the disease and how to prevent its occurrence. Enter Walter Reed, the chairman of the commission.

WALTER REED ENTERS THE FRAY

Walter Reed was born in Virginia in 1851 and earned his first medical degree in 1869 at the ripe old age of 17—at the time, the youngest ever to receive a medical degree from the University of Virginia. Continuing his studies at Bellevue Hospital in New York City, he received a second medical degree a year later, and his experience in the hospital clinics helped focus his interest on a career in public health.

After five years of interning and then serving in the Brooklyn Health Department, Reed passed the intense examination required to join the Army Medical Corps as a First Lieutenant. For the next 15 years Reed served in the Medical Corps, caring for soldiers. He also treated Native Americans during the short time he spent at a military post in Arizona. Reed's military career took him to postings all across the United States, but he also was able to continue his education, and acquired an expertise in physiology and bacteriology. His posting in Washington, DC, included conducting research at the Army Medical School, where he worked under the command of bacteriologist and Army Surgeon General Dr. George Sternberg, who was newly appointed in 1893.

George Sternberg had a remarkable career and was instrumental in the story of yellow fever. He had been an assistant surgeon for the Union Army during the Civil War and was captured in July 1861 at the First Battle of Bull Run, though he later escaped. The following year, during McClelland's campaign (if remaining stationary for months can be called a campaign), Sternberg was sickened by typhoid fever and spent the balance of the war years away from the battle fronts. Sternberg also contracted yellow fever during one of his military postings at Ft. Barrancas in the Florida panhandle, recovered, and published two papers on the disease in a medical journal.

Sternberg's experience with yellow fever landed him in Havana for service on the first Yellow Fever Board, along with two other doctors. In the early 1890s, Colonel George Sternberg vied for the position as army surgeon general and was selected over many more senior military doctors, based largely on his experience with both typhoid and yellow fever. Promoted to brigadier general, Surgeon General Sternberg faced a debilitating epidemic of typhoid fever then running rampant in US military camps and severely depleting its ranks. In the late 1890s, fresh recruits who had been mustered as a force to fight in Cuba faced their first battles in the stateside training camps—the enemies being unsanitary habits and impure water. More than one-fifth of the soldiers in training before deployment were sickened by typhoid fever and more than 1,500 died from the disease—more than would be killed in the ensuing fighting in Cuba. Sternberg called for the creation of a military commission, the Typhoid Board, to study and develop solutions to defeat typhoid fever.

The army's Typhoid Board was made up of three army surgeons and chaired by Sternberg's star pupil, Major Walter Reed. At the time, Major Reed was the army's leading expert on typhoid fever and its diagnosis. His expertise had been sharpened while serving as professor of bacteriology at the Army Medical School, created in 1893 by Surgeon General Sternberg. The Typhoid Board was charged with uncovering the underlying cause of the disease and proposing solutions. Board members traveled to numerous camps, made observations, and developed a solid picture of the disease and its transmission.

The board's thorough study—and equally thorough report—detailed the importance of sanitation in military camps, including locating camps away from low-lying areas and recommended building and maintaining latrines, improving cleanliness among the medical workers who were treating typhoid-sickened soldiers, and purifying water sources. The Typhoid Board had found solutions and provided them in a report to Surgeon General Sternberg. However, the recommendations were often ignored by base commanders, and typhoid continued unabated, hidden, the enemy within.

Typhoid was also one of the two major diseases faced by the occupying US force in Cuba—yellow fever being the other. Sternberg created the United States Army Yellow Fever Commission, known to more people as the "Reed Commission" after its leader, Major Walter Reed. Also serving on the commission were three other disease specialists, Drs. James Carroll, Jesse Lazear, and Arístides Agramonte. Although all four members were disease specialists, Lazear was the only one who had previously conducted research on mosquitos.

The Reed Commission set out to find the cause of yellow fever and develop recommendations for its prevention. One reason for the focus on yellow fever was the fear that soldiers returning from Cuba to the United States would bring it home with them, causing new epidemics. In fact, the Rough Riders were confined in quarantine on Long Island for six weeks when they returned, for that very reason.

BACK TO FINLAY

Reed Commission members focused on bacteria and what were termed "fomites"— objects such as the uniforms of sickened soldiers that had been soiled by human blood, excrement, and vomit. In regular reports, the Reed Commission had recommended cleaning and sanitary improvements, and they tested a variety of other solutions, all of which proved equally ineffective. Their focus on fomites was wrong. Their focus on bacteria was wrong. Their focus on sanitation also was wrong, but it provided a framework to help implement the eventual solution, once the causal agent was truly known. In a report on the commission's progress, Reed wrote, "at this stage of our investigation it seemed to me, and I so expressed the opinion to my colleagues, that the time had arrived when the plan of our work should be radically changed." They needed a different solution, and they needed to look back twenty years to find it.

Frustrated by their failures, commission members paid a visit to Carlos Finlay, to hear his 20-year-old ideas and reconsider his ignored theory that mosquitos spread yellow fever. Finlay had served on the earlier Yellow Fever Commission with General Sternberg, so he was known. Finlay's results were not always clear or positive, but the reason for those outcomes was not yet known. Over the previous 20 years, Finlay had performed experiments on more than 100 volunteers, inoculating them by allowing them to be bitten by mosquitos. Finlay's experiments were bold, considering that his ideas were viewed as foolish, if not totally wrong. However, his experiments were lacking in two important aspects. Finlay had not always exposed the mosquitos to infected patients during the period that they were infective—the first three days of illness—and he had used the infected mosquitos to inoculate volunteers too soon after the mosquitos had fed on infected patients, not considering the potential importance of an incubation period in humans.

Meanwhile, the "incubation period" piece of the yellow fever puzzle was solved in 1898 by careful observations and experiments conducted by Dr. Henry Rose Carter, an epidemiologist and quarantine officer for the US Public Health Service. At the time, Carter's findings were not known to the commission. Put this into perspective. The disease agent was not yet known to be a virus. It also was not yet known that mosquitos were the vector. For Carter to have solved the puzzle attests to how thoroughly he approached the problem.

Henry Rose Carter had been assigned to duty stations throughout the southern United States during the epidemics of the late 1800s, particularly during the 1878 and 1879 epidemics in New Orleans, Memphis, and the Mississippi River Delta. The environment in which Carter worked in rural Mississippi included isolated farmhouses where yellow fever patients lived and where he was able to track initial and subsequent infections. He discovered that individuals living in or visiting the homes of diseased patients either when or soon after the disease first occurred did not become infected. However, people who first visited two weeks later became infected. He later refined the conclusions from his careful observations, determining that an incubation period of 10 to 17 days after the initial human, infection was necessary for secondary cases to occur. However, he was unaware of the role played by *Aedes aegypti*.

Serendipity can be defined as finding something of value when it was unexpected or was not sought, but it matters only if one acts upon the finding. In one of the more serendipitous connections in solving the yellow fever puzzle, Carter was sent to Cuba in 1899, where he met commission member Dr. Jesse Lazear. After explaining his results to Lazear, the latter allegedly stated that "it spells a living host," meaning that a biotic vector was at play.

"ABSOLUTELY INCONTROVERTIBLE DEMONSTRATION"

Testing Finlay's theories was not so simple. Major Reed told Surgeon General Sternberg that, "only can experimentation on human beings serve to clear the field for further effective work." In a historic change from earlier medical research approaches, the commission took great care to ensure the experiments' volunteers understood what they were getting into. Earlier researchers used subjects that either had no choice (e.g., prisoners) or were vulnerable because they could not know the consequences of their service.

The studies in Cuba differed. The Yellow Fever Commission informed the volunteers (all males) about the experiments before initiating the research. In another important deviation from past studies, the commission compensated the volunteers monetarily. Volunteers were presented with written contracts that contained the terms of what was expected of them and how they would be compensated. Because some volunteers were Spanish or Cuban, they were given two contracts, one written in English, one written in Spanish. Ensuring volunteers understood the contract and that they had given their written, informed consent was of paramount importance, both for the well-being of the volunteers and to ensure the outcomes of the experiments were not flawed by volunteers not following protocols.

Reed and the other members of the commission gave the same amount of attention to the carefully considered experiments that they devised to test the theory. A series of elegant, but also controversial, experiments begun by Lazear provided the "absolutely incontrovertible demonstration" that Finlay had known would be needed to state conclusively that females of the common mosquitos *Aedes aegypti* were the carrier of the disease. In September 1900, without explanation to the other commission members, Reed left Havana for Washington, and the first experiments began. Using mosquitos hatched from eggs provided by Finlay, Lazear exposed a large batch of mosquitos to yellow fever patients so the mosquitos would take a blood meal. He then had the mosquitos feed on uninfected volunteers over the next fourteen days—different volunteers on different days. None who were bitten in days 1–10 became ill, but two volunteers who were bitten by infected mosquitos on days 12 and 14 were infected. The mosquitos whose bites caused the fever had fed on an infected patient at least 12 days before. Lazear's results lined up with those published by Henry Rose Carter—a minimum period was needed for incubation in an infected person. A mosquito feeding on an infected person before the incubation period concluded (more than ten days) could not transmit the disease; a mosquito feeding on an infected victim after the minimum incubation period was able to infect victims.

The earlier test subjects were not sickened because the disease had not incubated long enough in the mosquitos. The two volunteers who became infected included commission member Dr. James Carroll and a US soldier, William Dean. Although both Carroll and Dean were sickened by the fever, both recovered. Carroll, however, suffered damage to his heart and was weakened for the remainder of his life. Another "volunteer," was Jesse Lazear, who was bitten and became infected, but his infection proved to be fatal. Lazear showed symptoms on September 18 and was dead within a week. At the time, it was not known that he had intentionally allowed himself to be bitten, though careful examination of his laboratory notebooks proved nearly half a century later how he became infected.

The experiments showed Reed that Finlay's theory was correct—mosquitos had, in fact, transmitted the disease. There were two incubation periods—the time before a human could pass on the disease and the incubation period after a bite before the female mosquito could transmit the disease. Finlay's ideas were right and so the commission gained a victory in their war against the disease. But that victory came at a great cost in the loss of their colleague Dr. Jesse Lazear and the lifelong weakening of Dr. James Carroll. Yellow Fever was still winning the war.

"SILLIEST BEYOND COMPARE"—BUT CORRECT

Reed recognized that Finlay's mosquito theory was correct, but such recognition was not shared by the public. Following a presentation of the early results by Reed at a conference in October, a scathing newspaper editorial criticized the reported results as, "the silliest beyond compare . . . theories engendered by the mosquito hypothesis." Major Reed was determined that his colleague would not have died in vain. In an isolated area outside Havana, Reed ordered an experimental camp constructed to conduct the definitive experiments. In honor of their colleague who gave his life to find the answer to the yellow fever puzzle, the camp was named Camp Lazear.

For the first experiment, paid volunteers who had been screened from mosquitos were bitten by mosquitos that had previously bitten patients with yellow fever. Four volunteers became infected. The second experiment involved injecting blood from yellow fever patients. Three volunteers became infected. The third experiment was designed to test the fomites theory. Paid volunteers were assigned to stay in different, isolated buildings. In one building, screened to exclude mosquitos, volunteers were kept with fomites from yellow fever patients, to test the theory of bacterial contamination. Those volunteers slept on bed linens soiled with dried blood from patients who had died and also wore clothing belonging to those same fatally infected patients. None of the fomites volunteers became infected, putting to rest any remaining notions of fomites or bacteria being responsible for the disease.

Camp Lazear's other, sanitized, building contained two rooms: one with infected mosquitos and the second maintained free of mosquitos. Volunteers kept in the first room were bitten by the mosquitos and contracted yellow fever, but those in the empty room did not become infected. Fortunately, all the volunteers who became infected recovered. In a letter to his wife, Reed wrote that they "succeeded in demonstrating this mode of propagation of the disease, so that the most doubtful and skeptical must yield."

So it was that these experiments by the Reed Commission gave the "absolutely incontrovertible demonstration" of the role of mosquitos. The theory and experiments that had been labeled "silly and nonsensical" instead provided the key to the mystery that was yellow fever, that scourge that rode on the backs of the slaves who were sent to the New World to harvest the "white gold." The connections, the serendipity of the key personalities—Finlay, Carter, Sternberg, Reed, Carroll, Lazear—being the right people in the right place at the right time, and their work changed the trajectory of the history of insect-vectored diseases.

However, their work was not yet finished. They still had to translate the results into a plan that could be implemented to stop the disease. That would require yet another individual who would figure large in history: William Gorgas and his audacious plan that would lead to the successful completion of the Panama Canal. The silken thread formed the link—sugar to the Panama Canal, Cuba, Theodore Roosevelt, Carlos Finlay, and Walter Reed. The threads making the webs are there, and we can see them once we start to look for them.

CHAPTER 12

∽

William Crawford Gorgas and the Panama Canal

The Spanish-American War may have been the shortest of all the American wars, but it ranks near the top in the significant role played by insects. Its duration of just 114 days war pales in comparison to the 2,194 days of World War II and, especially, the 116 years of the Hundred Year War. The exact dates for that record-setting war are not known, and it was a series of conflicts rather than a continuous war, but it dragged on for more than 42,000 days. The relatively brief Spanish-American War was of course significant for reasons other than insects, in particular the changes it brought about in the world's power structure. Spain declared war against the United States on April 24, 1898. This was followed the next day by a declaration of war by the United States against Spain, though the date of the declaration was made retroactive to April 21—nothing like beating them to the punch!

With battle fronts in Cuba and half a world away in the Philippines, both military forces were stretched thin. The fighting in the Philippines is known best—if known at all—for the Battle of Manila Bay, with that naval encounter being remembered mostly for two things: Commodore George Dewey's famous order, "You may fire when you are ready, Gridley," given to Captain Charles Gridley, who then began the battle. And Dewey's parlay of this popular victory into his candidacy for president of the United States in the 1900 election—though he eventually withdrew from the race and endorsed William McKinley and his vice-president, Theodore Roosevelt, famed for his role in the war in Cuba. But we are jumping ahead again.

Fighting on the war's primary front in Cuba lasted only a few months, with even less time spent in actual combat. An armistice was signed on August 12, 1898, with the Treaty of Paris signed on December 10 of that year. The treaty resulted in Spain turning over the territories of Puerto Rico and Guam to the United States and transferring sovereignty of the Philippines to the United States for the sum of $20 million. The war and its conclusion resulted in Spain withdrawing from the world stage and

The Silken Thread. Robert N. Wiedenmann and J. Ray Fisher, Oxford University Press. © Oxford University Press 2021.
DOI: 10.1093/oso/9780197555583.003.0012

the end of its long colonial history. The United States, bolstered by the military victory, assumed Spain's place as a key player in world geopolitics, a position that would continue to grow over the next century.

Insects played a role, both in the Spanish-American War and in the positioning of the United States to become a world power.

Despite the armistice, a battle continued in Cuba, though against unseen and as-yet-unknown enemies—the causal agents of yellow fever and malaria. Both diseases, accompanied by typhoid fever, raged on during the US occupation that followed the armistice. Colonel Theodore Roosevelt and his Rough Riders had returned to the United States, where they were quarantined for six weeks to avoid bringing the fever onto home ground with them, and to keep the heroes safe. However, the large US occupation force remained in Cuba and squarely in harm's way.

RIGHT PERSON FOR THE JOB

Leading the charge against the diseases ravaging US troops was Major William Crawford Gorgas, the army's chief sanitary officer in Havana (Figure 12.1). Gorgas is known for his later triumph in the battle against yellow fever in Panama, but what he learned and implemented in Havana during the occupation set the stage for the ultimate victory in Panama. The military governor of Cuba, Major General Leonard Wood, put Gorgas in charge of fighting the diseases after the war ended. In Gorgas, the army had the right person for the job.

William Gorgas had a storied life, a life that led him to Havana and, later, to Panama. He was born in 1856 in southern Alabama. His grandfather, John Gayle, had previously been governor of Alabama. His family moved around during William's childhood, because his father, Lieutenant Josiah Gorgas, was in charge of several US Army arsenals prior to the Civil War. A year after the war ended, the family relocated to Alabama, where much of William's youth was spent in outdoor activities while he dreamed of making a career in the military. His father became headmaster at the University of the South, and William joined his father and enrolled there, graduating in 1875.

Pursuing his dream of a military career, Gorgas sought admission to West Point, but he was not accepted. Instead, he enrolled at the Bellevue Hospital Medical College in New York City and graduated in 1879. In June 1880, he joined the US Army Medical Corps as a first lieutenant and assistant surgeon and was posted to Ft. Brown, in Brownsville, Texas. The posting in south Texas would change his life and the trajectory of his acclaimed medical career.

Ironically, the game-changer was contracting yellow fever. Fortunately, it was not a serious bout, and the infection turned out to be beneficial for him both personally and professionally. While recuperating from the fever, he met his wife-to-be, who was also recovering from yellow fever. His recovery conferred lifelong immunity, enabling postings to places where outbreaks occurred, such as Ft. Barrancas, Florida. Even more, his immunity led to postings at two key locations where his vision and dogged efforts would change forever the fight against yellow fever.

Figure 12.1 US Army Surgeon General William Crawford Gorgas in 1914. Gorgas was responsible for eliminating yellow fever in the Panama Canal Zone.
Source: Photograph by Harris and Ewing, courtesy the Library of Congress

Major Gorgas was sent to Havana as Chief Sanitary Officer in July 1898, where he was positioned to fight several diseases afflicting the occupation troops. The crowded city was the key to controlling the diseases, for reasons beyond it being the occupation force's command center. Passengers arriving at the port of Havana could be screened, controlled, and monitored. Because Major Gorgas believed yellow fever was caused by poor sanitation and because he was chief sanitary officer, he focused on improving sanitary conditions. His focused efforts helped control typhoid fever and dysentery in Havana, both of which spread more easily in unsanitary conditions. However, sanitation had little impact on yellow fever, which returned to Havana in the summer of 1899. The enigmatic disease defied all attempts to prevent it—until the Reed Commission report pointed out that the agent spreading yellow fever was literally right before their eyes, noticed, but not recognized. A mosquito.

The idea of a tiny insect being the vector of the disease was not universally accepted, even after the experiments conducted by Major Walter Reed and others clearly demonstrated its role. The idea pre-dated the report—the Cuban physician, Dr. Carlos Finlay, had been saying for two decades that mosquitos transmitted the disease. But no one paid any heed to Finlay's idea until Reed's experiments convincingly confirmed what Finlay had been saying.

Confirmation that common domestic mosquitos, *Aedes aegypti*, were the agent of yellow fever's spread was a double-edged sword. Identification of the mosquitos and knowledge about their life history gave the military a target to fight. However, this unconventional target required unconventional tactics, different to any tactics used before in any battle. Under orders from Military Governor Leonard Wood, Gorgas first pursued inoculations to confer resistance to the disease, using volunteers in his experiments. Unfortunately, the trials proved fatal to many of the test volunteers and the experiments were cancelled. Gorgas then pursued the only option that remained—eradicating the mosquitos.

ERADICATE MOSQUITOS?

Anyone who has traveled to tropical countries knows how ludicrous an idea eradicating mosquitos was. If the idea of mosquitos as disease vectors was considered ridiculous, the suggestion of eradicating them was even more preposterous. Mosquitos just were. Daytime mosquitos. Nighttime mosquitos. Mosquitos active only at dusk and dawn. They were everywhere and anywhere, all the time, especially in tropical countries. Hot and humid Cuba, with a rainy season that lasted nine months, provided the ideal habitat for mosquitos.

Eradicate mosquitos? The military was prepared to fight by infantry attacks or cavalry charges, as they had in the recent battles on Cuba's San Juan Heights. But the Gatling guns integral to victory in that battle, or the new 0.30–0.06 Springfield rifles issued to infantry troops fighting in Santiago, were of no use in the battle against mosquitos. Yet, the battle plan drawn up by Major Gorgas was no less detailed than plans for conventional war. Gorgas brought to bear his keen logic, experience with the disease, and the full power of the military to the fight against mosquitos.

In a paper published in 1901, Gorgas presented the details of his plan, beginning with two key assumptions that are now accepted as necessary for yellow fever's spread: (1) presence of a non-immune person capable of being infected; and (2) infected mosquitos that could transfer the disease to non-immune people. Those two assumptions determined the two major battle fronts in Gorgas's plan of attack. At that time, more than 200,000 people who were immune to yellow fever lived in Havana along with roughly 30,000 "non-immunes." Removal of the non-immune population was the obvious solution, but even the military were limited in what they could do. And many of the non-immunes were the members of the US occupational force, so removing them would hardly constitute "occupation."

Unlike removing susceptible occupation troops and support staff from harm's way, increased scrutiny of everyone arriving at the port of Havana was possible. On disembarking, arriving passengers were separated: those headed any place in Cuba other than Havana were immediately sent to the countryside without going through the city. Arriving passengers going to Havana were inspected at that time and again later in follow-up visits. Anyone who was found ill in the follow-up inspections was hospitalized for two months(!)—that is, if they didn't die—with the costs paid by the military.

In rural towns, the non-immunes were sent to nearby camps to keep them healthy. Despite this tactic's similarity to the recent rounding up of locals by the Spanish, the temporary removal ordered by the US Army was seen as a way for the occupying troops to help the local community. This approach had been used successfully to stop epidemics such as typhoid fever and dysentery at military posts. Although the non-immunes were not isolated from mosquitos, their isolation from other people, especially in other towns, denied the mosquitos access to new healthy, non-immune people. The first battle in Major Gorgas's plan was won, much as a military could emerge victorious by cutting off an enemy's supply line.

CIRCLING THE ENEMY

Gorgas's second plan in the war against mosquitoes was a multi-pronged attack directed against them. Major Gorgas had two small divisions of soldiers placed at his disposal. The first division was divided into three-person teams, each consisting of a foreman to inspect houses and two laborers to do the work needed to eliminate breeding sites for mosquitos. Each team was assigned to one of the 13 city's districts. The teams visited every house in their assigned district once every three weeks, making modifications as needed. They used oil to form a film over any standing water to kill mosquito larvae, consuming some 8,000 liters of oil per month. They also inspected rain barrels to ensure they were mosquito-proof, as had been ordered, and destroyed any that were not in compliance. Yellow fever mosquitos habitually live near their human hosts, so denying them places to deposit their eggs removed them from the vicinity of their hosts.

The second division of 50 men was assigned to clean ditches and streams in and around Havana, draining low places that held water suitable for mosquitos to breed or covering the water with a film of oil if drainage was not possible. Both tactics served to reduce the numbers of mosquitos by eliminating places for them to lay eggs.

The next prong of the attack was to prevent mosquitos from becoming infected, thereby eliminating their ability to transmit the virus to new hosts. This strategy required making sure that patients suffering from the fever were not bitten again by mosquitos. Dwellings holding infected patients had screens placed on the windows and doors, and the doors guarded to ensure only designated immune persons could enter. Hospital wards were also screened. A squad deployed to the house of an infected patient sealed the house and (after the patient was removed) burned pyrethrum powder at rates of about 1 kilogram per 30 cubic meters to kill any mosquitos that remained. Adjacent houses were treated in the same way.

Finally, all the roads that led from the countryside to Havana were guarded and all people traveling to Havana were scrutinized. Every individual needed to provide inspectors their name and address in Havana, so that an inspector could visit them after three and six days to verify their continued health and, if they became ill, isolate them.

The extensive program was nothing if not thorough, and it had all the hallmarks of a well-designed military battle plan. Yellow Jack was no match for Major Gorgas. Total yellow fever deaths in Havana during the "fever season" from April through the end of August in 1900 were just 89. During the same period in 1901, with Gorgas's battle plan fully implemented, the number of deaths was three. Three! Realize that in 1901 there were more non-immunes in Havana than at any time in the previous decade, and deaths even as recently as 1897 had exceeded 600.

Continuing, in 1902 there were no reported cases, showing just how effective Gorgas's plan was against the mosquitos and the disease. Another benefit from the thorough, multi-pronged program against yellow fever mosquitos was that deaths from malaria—spread by different mosquitos (*Anopheles*) with very different biology—also dropped dramatically. Yellow fever was permanently gone from Havana. Major William Gorgas had won, defeating the tiny enemy in a battle nearly 400 years in the making.

CONNECTING THE OCEANS

While the battles of both the Spanish-American War and against yellow fever played out in Cuba, the fever was wreaking havoc in Central America. That battleground was located in the isthmus of Panama, the long and narrow country that joins the continents of North and South America. At its narrowest, only 60 kilometers separate the Atlantic and Pacific Oceans, but that separation required any ocean-going passenger traffic or freight shipments to make the long journey around South America. Sailing from one ocean to another was always challenging, not solely because of the distance to Cape Horn at the far southern tip of South America, but also because of the ferocity of ship-sinking storms that raked the Cape.

A canal that could join the oceans had been proposed some 300 years earlier, in the late 1500s, when Spain's King Charles I ordered a survey of a route that might join the two bodies of water. At the time, constructing a canal through the jungles and across mountains was not only impossible, but impossible to even consider. But that would change at the time of the war in Cuba.

Ideas about marine routes to connect seas or oceans are at least 4,000 years old, with small canals linking the Nile River with the Red Sea used by trading boats as early as 2,000 BCE. There was great interest in a water route that connected the Red Sea with the Mediterranean Sea but, even as late as the early 1800s, it was commonly believed that the two seas were at different altitudes. The logic behind that belief isn't quite clear. Misinformation and mistaken beliefs occur in every age. A survey conducted in the 1830s finally demonstrated that the two seas were not, as had been thought, at different altitudes, and these definitive results enabled planning for a sea-level canal.

Numerous individuals figured in the planning and construction of the Suez Canal linking the Red and Mediterranean Seas. Here, we mention only the French diplomat, Ferdinand de Lesseps, who was granted permission to create a company to construct the canal. Construction began in 1859 on the Suez Canal, and ten years and $100 million were required to complete it—a sum more than twice the original estimated costs.

While the Suez Canal was under construction, another, equally important construction project began half a world away. Whatever reason the group of New York businessmen had for wanting to build a railroad across Panama, their timing was fortuitous. After negotiations with the government of New Granada (later Colombia), which controlled Panama at the time, the Panama Railroad Company was formed in 1845. Railroad owners began shaping plans to build what would become the world's first transcontinental railroad, even though "transcontinental" in this case meant traveling less than 80 kilometers. What made their timing fortuitous was gold. Not the "white gold" of Caribbean sugar plantations, but honest-to-goodness gold, discovered in California.

The oft-cited "first discovery" of gold in January 1848 by James W. Marshall, a carpenter constructing Sutter's Mill, was momentous, but it was not the first. Gold had been discovered at least six years earlier some 600 kilometers to the south in the Santa Clarita Valley, just north of present-day Los Angeles, while California was still a territory of Mexico.

Reports of gold mining earlier than 1842 are somewhat murky, but what is known for certain is that roughly 600 *kilograms* of gold was mined from Placerita Canyon in the Santa Clarita Valley between 1842 and the 1848 discovery at Sutter's Mill. That amount would qualify as not trivial, though hardly a fleck in the pan compared to the 3,350 *metric tons* of gold extracted at Sutter's Mill from 1848 to 1853. Marshall's discovery at Sutter's Mill actually occurred before the signature of the Treaty of Hidalgo, which ended the Mexican-American War and ceded California to the United States. Did the United States give Mexico "its" gold? Want to guess?

The reports of gold being found made their way from the Sierra Nevada Mountains of northern California to the eastern United States by September of that year, triggering a worldwide "gold rush" of people heading for California. The Panama Railroad was completed in 1855, so its trains were not able to carry the initial wave of "49ers" on their journey to riches. However, a wave of late-comers headed to California from the eastern United States availed themselves of the railroad. The tripartite journey of sailing to Panama and then on to California, with the middle leg of the trek on the railroad, shortened their trip considerably. But to do so carried a steep price—as much as $25 in gold for a transcontinental journey of less than 80 kilometers.

Even though the Panama Railroad may have been late in the game for the gold rush, once the railroad connected the two oceans, it found considerable and important use. During the US Civil War, the railroad was used to transport soldiers and material from east to west and vice versa. Similar overland transport was not yet available in the United States, where the transcontinental railroad would not be completed until 1869, four years after the Civil War ended. The success and importance of the Panama Railroad only amplified interest in a canal to connect the two oceans.

PANAMA IS NOT THE SUEZ

Following the successful construction of the Suez Canal (despite the huge cost over-run and financial trouble), France set about constructing a sea-level canal across the isthmus of Panama, under the leadership of the same Ferdinand de Lesseps. The project had its ceremonial kick-off in 1880, but actual construction began in earnest only a year later, when de Lesseps's French Canal Company bought controlling shares in the Panama Railroad Company. The railroad was seen as an integral resource that would be needed to haul supplies and workers during construction, as well as to carry away massive amounts of soil and rock from excavations.

Almost immediately, the project ran into significant challenges. Unlike the Suez Canal, which was constructed in a relatively flat desert, Panama's mountainous land was soaked in never-ending rains that generated significant landslides. Also unlike Suez, Panama was home to clouds of mosquitos that were then known only for making life miserable, not for being deadly carriers of yellow fever and malaria. The French had hired 30,000 workers from the West Indies to cut through the jungle, blast through rock, and excavate the canal. The large number of workers was due, in part, to the many who were incapacitated by accidents or the ever-present diseases.

A French engineer, Philippe Bunau-Varilla, took over as construction manager in 1894 but, too late, de Lesseps realized that the planned sea-level canal would require too many resources—both financial and human. He changed plans to pursue a canal with locks to carry ships over the mountains, using a design from Gustave Eiffel—yes, *that* Eiffel. But investors grew weary of the slow progress, the changes in plans, and the continued plea for more funds. De Lesseps's charisma and reputation as the master of the Suez were finally played out.

Despite de Lesseps's promises, Eiffel's design, the infusion of cash, and the large number of workers, the project ended ignominiously in 1889 when the company led by de Lesseps was declared bankrupt. The amount of money spent by the French in Panama is in dispute, with figures ranging from more than $280 million to nearly double that. The true figure is likely nearer the high end.

The liquidation of the Canal Company unmasked financial impropriety at its worst—lavish spending of money that never made its way to Panama, extravagances in Panama, and millions of dollars (US) paid in bribes to the press, banks, and more than 500 members of the French parliament. That was the financial cost. The human cost was far greater. Financial mismanagement and corruption were blamed for the project's failure, but the major causes were the deadly diseases—yellow fever and malaria.

Exact numbers of fatalities during the French canal-building efforts are not known, but upwards of 25,000 West Indian workers, French engineers, and their families died, most of them from fevers. In one of the most sobering examples, the director general of the French company, Jules Dingler, moved to Panama in 1883 along with his wife, son, daughter and her fiancé. In just a few months, his daughter and son died of yellow fever, followed by his daughter's fiancé and, within a year, his wife. Dingler never occupied the extravagant mansion and estate constructed to house his family and his collection of horses. His entire family gone and his dreams shattered, Dingler departed Panama for good and returned to France. Yellow fever had gotten personal and, as was often the case, emerged the victor.

PICKING UP THE PIECES

The Spanish-American War made even more clear the need for a connection between the Pacific and Atlantic Oceans for use by the US Navy. The turn of the new century brought a treaty that gave the United States rights to build and operate a canal in Panama, at the time still part of Colombia. That geopolitical status changed in 1903 when Panamanians declared their independence from Colombia, the declaration enabled silently by the presence offshore of an American warship.

The United States wasted no time in drafting and signing a new treaty, receiving a narrow strip of land across the isthmus in exchange for $10 million. In addition, the remaining French equipment and other assets were purchased in 1904 for $40 million. With a treaty in hand, payments made for permission to develop and operate a canal, and equipment and other assets in place, the United States set out to build a canal, under the direction of the Isthmian Canal Commission and with the full support of President Theodore Roosevelt.

Roosevelt appointed John Findley Wallace as first chief engineer on the canal project. Upon his arrival in Panama, Wallace found that the purchased equipment from France was not quite what had been advertised: much was unusable, unsalvageable, and even overgrown by tropical vines. The rail line was in disrepair, buildings abandoned, remaining minimal infrastructure poor, and sanitary conditions even worse. Despite the lack of operating equipment and the poor conditions, Wallace directed a crew to continue the work begun by the French, to blast through the mountains at a site called Culebra Cut. Almost immediately, the US-led effort ran headlong into the same difficulties that had been faced by the French. Conditions in Panama were miserable: few roads, little housing for workers, unsanitary conditions, and disease that ran rampant, sickening and killing workers. Wallace, not familiar with tropical conditions, and certainly not familiar with directing a project with all the challenges faced in Panama, resigned after only one year, fearing for his own personal health and safety.

Wallace's departure turned out to be a blessing in disguise. Roosevelt named John F. Stevens to replace Wallace as chief engineer. Stevens, having had considerable experience in railroad building, realized that the lacking infrastructure needed to be restored before the true canal work could proceed. For a full year, the workers excavated no rock at Culebra Cut, but instead were tasked with building sufficient and safe housing, infrastructure to provide clean water, and sanitary sewers, to create living conditions improved enough to attract—and retain—workers and their families. In that year, Stevens's crew built more than 1,200 homes, administration buildings, and hospitals. Another 1,200 French-constructed buildings were rebuilt and restored. When work resumed on the blasting of the mountains, at least the living conditions were tolerable.

RE-ENTER COLONEL WILLIAM GORGAS

Stevens's wise decisions and strong leadership set a course for the canal construction to proceed apace. Almost certainly, the wisest decision Stevens made was to listen to and support the plan proposed by the Canal Zone's chief sanitary engineer—Colonel William Crawford Gorgas, who had earned a promotion to colonel and assistant surgeon general for his success against yellow fever in Havana.

Colonel Gorgas sought the assignment in Panama as soon as he learned of the US-led project, and so he went to Panama. The year was 1904. That point in time is important, because the recent experiments conducted by Colonel Walter Reed and his colleagues in Cuba had established mosquitos as the vectors of yellow fever and malaria, though their findings were still largely not accepted by the medical establishment of the day. Having seen first-hand the role played by yellow fever mosquitos in Cuba, Gorgas developed and executed a battle plan for Panama based on the one that defeated the mosquitos and disease in Cuba.

The thorough battle plan enacted and completed by Gorgas in Cuba provided a template for controlling mosquitos, and thus, yellow fever, in Panama. It was based on evidence, not long-held beliefs. Despite his plan being based on evidence from military doctors, Gorgas did not have the full support of the military and its leaders for his multi-pronged attacks. Military leaders did not believe their own doctors! Nor were the Isthmian Canal Commissioners even sympathetic, let alone supportive of his plan. Despite the success in Havana, Colonel Gorgas was mocked for his belief in the controversial "mosquito theory" of disease transmission, because "experts" in the States knew that yellow fever was caused by miasma—bad air or soil-borne toxins. He was even advised by one of the commissioners to get the idea out of his head.

Colonel Gorgas had a plan, but he needed money to implement it. Big money. He went to Washington, DC, in late 1904 to try to get his plan funded. The price tag for his plan was steep—$1 million to *start* the project. He was rebuffed and roundly criticized for such an outlandish request for money to throw at an unproven theory.

Let's pause momentarily and put this request into context. First, the entire budget for the US government in 1904 was $584 million, and Colonel Gorgas was requesting $1 million. Second, he wanted that million to kill mosquitos in the faraway country of Panama. And third, this request was to start the project, and who knows how much the final price tag would be? In Panama? Is it any wonder he was greeted with such skepticism?

The following spring, an outbreak of yellow fever brought death and dismay to Panama. Project managers had trouble attracting workers to sign up to work in the disease-ridden country and, without workers, the canal project was in danger of failing. That outbreak turned into another positive in Gorgas's career. His request for funding was still resting with Roosevelt, with no decision and no action yet taken. With news of the outbreak, Roosevelt's close friend and family physician Dr. Alexander Lambert spoke candidly with the president and laid out the options: Revert to old, ineffective methods and the canal will fail, just as the French effort had failed. Follow Gorgas and his plan and you will get your canal. Talk about serendipity! Right person, right place, right time. Dr. Lambert may have been the unsung hero of the canal project. Roosevelt threw his support behind the request and Congress provided the funding.

MOSQUITOS MEET THEIR MATCH

As he had in Cuba, Gorgas had developed a multi-pronged plan for ridding the Canal Zone of yellow fever, but the plan for Panama was even more far-reaching and impactful. With funding backed by Roosevelt and the support of Director Stevens, Gorgas divided the Canal Zone into districts, with houses and buildings in each district inspected by teams. If the inspectors found standing water—potential mosquito breeding habitat—in, on, or around a structure, it was corrected by the team's carpenters. And they were thorough! Even containers of holy water in the churches were emptied and cleaned.

The second prong of Gorgas's attack plan included rural areas. Swamps and wetlands in and near the Canal Zone were drained and filled by the "Gorgas gangs." Standing water pools and streams were treated with oil and larvicide to kill any developing mosquito larvae; this part of the project used about 2.8 million liters of oil and 500,000 liters of larvicide. Brush was cut over a wide area. Gorgas also developed and implemented a domestic water system to provide clean, safe water to residents. Clean water piped from a nearby lake eliminated the need for the rain barrels that had collected water but also bred mosquitos.

The sanitation plan Gorgas enacted was far-reaching. It included a modern sewer system, introduced indoor toilets, and established regular garbage collection. Local health departments were created and were responsible for burying or cremating the dead, and ended the practice of "renting" and reusing graves. They also distributed a ton of quinine per year to combat malaria. The homes of all people who contracted yellow fever were fumigated with pyrethrum or sulfur. Once the success of fumigation was recognized, that tactic was extended to all the buildings in the terminal city of Panama, which required using the *entire* supply of sulfur and pyrethrum held by the United States in 1905.

The thorough efforts of Gorgas's teams yielded resounding successes, but his battle plan contained still more tactics. A critical part tackled disease (including malaria) by quarantining infected individuals in isolation cages to prevent mosquitos from biting and transmitting the diseases to new patients. Building windows were screened. Workers slept in verandas that were screened to keep away night-biting mosquitos, which were the ones spreading malaria. Malaria at that time was so rampant that, in the other Panamanian terminal city Colón, one of every six residents suffered from malaria every week.

Gorgas even influenced the treatment of yellow fever. Although there was no specific prescribed treatment, Colonel Gorgas claimed better survival rates in his Panamanian hospital (Figure 12.2) when the treatment consisted of complete bed rest and nothing given by mouth except sips of champagne. Champagne as a treatment? No wonder Gorgas and his project budget were questioned! The same disease-ridding

Figure 12.2 Ancon Hospital grounds, Panama Canal Zone, ca. 1910.
Source: Wellcome Images/CC BY 4.0

efforts were applied to malaria, but that fever was less tractable because of different mosquito biology. Still, the efforts against yellow fever also significantly reduced cases of malaria.

Colonel Gorgas was not just fighting mosquitos. The bigger foes were his military commanders and the Isthmian Canal Commissioners. The French had failed because of disease, yet the measures taken by the Americans in charge of the canal's construction were pushing their effort toward the same dismal outcome. Admiral John C. Walker, Head of the First Canal Commission, was determined not to repeat the mistake the French made with graft and extravagance (such as "Champagne treatments"). Walker was convinced that improving sanitation was plainly and simply "extravagance."

Environmental conditions in Panama were more complex than they had been in Cuba, so successes were realized more slowly, which was interpreted by impatient commissioners as "failure." All the while, Gorgas was the focus of a campaign to discredit him and his work, with the goal of seeing him replaced by someone with "more practical views," which meant someone who didn't focus on mosquitos. Gorgas soldiered on.

Admiral Walker was replaced on a new commission headed by Theodore Shonts, who could have spelled the demise of the canal project. Shonts right away wondered why the primary interest of the sanitation service was eliminating mosquitos. His commission openly called for Gorgas to be removed and for cutting the "outlandish" expenses of sanitation.

Figure 12.3 President Theodore Roosevelt running an American steam-shovel, Panama Canal, 1906.
Source: Underwood & Underwood

However, in his zeal to cut costs and set the project back on the "right" path, Shonts went a little too far. His vocal opposition of Gorgas reached President Roosevelt—the same President Roosevelt who had been persuaded by Gorgas and Dr. Lambert to use all-encompassing methods to fight the disease. Roosevelt reiterated his support for Gorgas and his plan.

Whether to further demonstrate his support or for other reasons, Roosevelt traveled to Panama for three days in November 1906, which was not only the first time a sitting president traveled outside the United States, but it may also have been the first historical "photo op"—Roosevelt was photographed sitting on and operating the huge, 95-ton Bucyrus steam shovel (Figure 12.3). Bully! Gorgas stepped up his efforts, sticking with his battle plan.

THE PREPOSTEROUS AND IMPOSSIBLE PLAN WORKED

Despite the nay-sayers, the threats of dismissal, and the denigration of Gorgas's work and service, the plan worked. Yellow fever cases and deaths dropped dramatically. By August 1905, the numbers of new cases of yellow fever had fallen by half;

in September, the monthly tally of new cases was seven, with the last death from yellow fever recorded in November 1905. Work on the canal, building the series of great locks (Figure 12.4) to carry ships across the isthmus, proceeded without the accompanying fever. The unthinkable had happened—the preposterous and impossible battle plan led by Colonel Gorgas succeeded, and that plan resulted in the canal's successful completion. One can imagine Colonel Gorgas was congratulated warmly and effusively by Walker and Shonts. Or not.

One life that wasn't saved was a casualty not of yellow fever, but of more common, mundane causes. Appendicitis, though regularly encountered in the early 20th Century, was every bit as deadly as yellow fever, especially when treatment was not provided soon enough to prevent peritonitis. Colonel Walter Reed was 51 years old when his appendix ruptured. The peritonitis that followed was too much for attending surgeons to turn back. Walter Reed died on November 22, 1902, having seen the results of his work succeed in Havana, but before the final defeat of yellow fever mosquitos in Panama.

Collateral benefits of Gorgas's plan included major reductions in cases of malaria. Malaria deaths dropped 10-fold in three years—it was not eradicated, but occurred at levels low enough to allow the work on the canal to proceed safely. Gorgas's successful

PEDRO MIGUEL LOCKS. LOOKING SOUTH FROM EAST BANK. SEPTEMBER 14, 1910.

Figure 12.4 Construction on the Pedro Miguel Locks, part of the Panama Canal.
Source: Wellcome Images/CC BY 4.0

battle plan has been credited with saving the lives of at least 70,000 people who went on to work in the safe and sanitary Canal Zone.

Oh. One more collateral benefit. The Panama Canal was built. Its completion in 1914 marked the most significant modern engineering marvel then known to humankind. Ten years of project leadership from the United States came at a financial cost of $352 million. The human costs were also great. About 5,000 deaths during the construction were attributable to diseases, mostly malaria and yellow fever. Ten additional years of French efforts, cost at least $280 million, likely more, and more than 25,000 deaths, again, mostly from yellow fever and malaria.

And what of the key players? Most readers have heard of Walter Reed. It is unlikely that the name Alexander Lambert has been encountered in casual reading or while completing a crossword puzzle. The names of the various commissioners are all but forgotten, and perhaps rightly so. But William Crawford Gorgas? Readers most likely do not know of Gorgas or his successes; successes that accompanied him wherever he went. He was elected president of the American Medical Association in 1908 and named surgeon general in 1914 by President Woodrow Wilson, and reached the rank of Brigadier General. He was the world's foremost expert on sanitation and disease. While en route to Africa in 1920 to consult on disease control, he suffered a stroke and was hospitalized in England. Britain's King George V visited his hospital room and knighted Dr. Gorgas on his deathbed.

The largest military medical complex in the United States is named for Walter Reed. Gorgas? His name appears on the London School for Hygiene and Tropical Medicine, on a small research building in the Canal Zone, and on the original hospital at Fort Brown, which is now a building on the campus of Texas Southmost College. Reed is credited with solving the riddle that was yellow fever, though the credit should be shared with (or given to) the Cuban doctor, Carlos Finlay. Gorgas, against all odds, put Reed's and Finlay's findings to work and rid Havana and the Panama Canal Zone of yellow fever. The Panama Canal made it possible for the United States to become a 20th Century global military and industrial superpower. All Dr. William Gorgas did was make the canal a reality. By defeating a mosquito.

SECTION 5

Western Honey Bee

—6000 BCE	Honey hunting is depicted in cave paintings in Spain
—2500 BCE	Beekeeping is recorded in Egypt (begins earlier)
—1000 BCE	Beekeeping is evident in ruins from Tel Rehov (modern Israel)
—550 BCE	Beekeeping is taken to Persia
—63 BCE	Bees are used as weapons in the Third Mithridatic War
—1191 CE	Richard I uses bees and catapults in the Siege of Acre
—1622 CE	Western honey bees are taken to the North American colonies
—1914 CE	Bees defeat British troops in the Battle of Tanga in East Africa
—1956 CE	African honey bees are taken to Brazil, from where they escape
—1973 CE	Karl von Frisch wins the Nobel Prize in Physiology and Medicine

CHAPTER 13

༄

Six-Legged Livestock

Like us, flowering plants have sex. Unlike us, they need help from other species to be successful at it.[1]

We may rarely think about it, but most familiar plants do indeed have sex. Sexual reproduction involves the transfer of gametes to produce a new organism and is the mode of reproduction for most multicellular species. Gametes are the cells that contain a single set of an organism's chromosomes (haploid), which, when united with the gamete of a member of the opposite sex, produces offspring with two sets of chromosomes (diploid). For both plants and animals, there are two sexes (male and female), defined by the size of the gamete (smaller and larger, respectively).[2] The successful transfer of gametes—the bane of teenagers in the back seat of a car—is an absolute requirement for organisms that reproduce sexually. For plants, the gametes are transferred in structures known as pollen.

POLLEN AND POLLINATION

Apologies for skipping stones across the intricacies of plant reproduction. Our focus is on pollen and its transfer, pollination, which is how most familiar plants reproduce. Individual grains of pollen are small, ranging in size from about 7 micrometers (a micrometer is 1/1000 of a millimeter) to more than 700 micrometers in diameter. The size, shape, and surface features of pollen are specific to each species of plant, and images of pollen from modern microscope techniques (e.g., colored Scanning Electron Microscopy) are dazzling. Pollen has to get from the male to the female flower and there are two primary means by which that occurs. Pollen of some plants is carried by the wind (these are called anemophilic: anemo=wind, philic=loving) whereas other pollen is transported by animals, mostly by insects (called entomophilic: entomo=insect, philic=loving).

The Silken Thread. Robert N. Wiedenmann and J. Ray Fisher, Oxford University Press. © Oxford University Press 2021. DOI: 10.1093/oso/9780197555583.003.0013

Maybe you have an allergy to pollen, so you are ready to curse all pollen. Not so fast! You are most likely not allergic to all pollen, but to anemophilic pollen, whose grains are either small or have projections that act like wings, and the pollen is borne by the wind. For example, most grasses (e.g., corn) produce wind-blown pollens that are common allergens. In general, plants that reproduce by wind pollination[3] produce a lot of small pollen grains that contain the gametes and little else; they are devoid of nutrition. In contrast, entomophilic pollen, not reliant on the wind, is generally larger and sticky because this type of pollen is actively carried by animals, especially insects. The larger entomophilic pollen often is nutritious and fed on by insects and other pollinating animals, such as bats and birds. Most plants that produce entomophilic pollen have flowers that produce nectar. Flowering plants also add color to our landscapes and yield the fruits and vegetables we eat, and most of our focus is on these flowering plants.

Who exactly are the pollinators? Most are insects, but we'll get to them in a moment. Bats, birds, and even humans pollinate plants. Although most bats feed on insects, several prominent groups of bats are pollinators, including the leaf-nosed bats and fruit bats. At least 500 species of plants are pollinated by bats; not surprisingly, most of those plants have flowers that open at night. Birds are even more numerous as pollinators, with nearly 20% of all bird species transferring the pollen of thousands of plant species. Groups such as sunbirds, honeyeaters, orioles, and some parrots are pollinators. The most prominent are the hummingbirds, who brush against sticky pollen as they feed on a plant's nectar, transferring pollen when they move to the next flower to feed. There are more than 300 species of hummingbirds, and some are highly specialized. For example, sword-billed hummingbirds (*Ensifera ensifera*) live at high-elevation in the Andes; their bills, which are more than 10 cm long, make them uniquely able to reach pollen of flowers with long corollas, such as *Passiflora* and *Datura*.

But most pollinators, by far, are insects, and most plants utilize insects for pollination. In this chapter, we focus on honey bees, but it is important to note that not only are other bees important pollinators, so too are many other insects. Bees were not the first pollinators. Flowering plants evolved about 130 million years ago, and the bees are thought to have arisen about 100 million years ago—that leaves 30 million years of pollination by other insects. A variety of insects are pollinators, especially beetles and flies. Most beetles are generalist pollinators, meaning they will visit many plant species. In many cases, generalist pollinators are crucial. A rare plant, for example, may not be able to attract or support specialist pollinators; without generalists, the plant could not reproduce. Many flies are generalists, but specialists abound as well. For example, cacao—the tropical tree whose fruit is processed to become chocolate—is pollinated by tiny flies called chocolate midges (*Forcipomyia* and *Euprojoannisia*). So far, they are the only insects known to be able to pollinate the small flowers of the cacao tree. If you like chocolate, thank the specialist midge that makes it possible.

The dynamic between plants and insect pollinators is one of mutual benefit. Many flowering plants entice insects by producing heavy, nutritious pollen and nectar, and the plants advertise their presence with showy flower parts and scents, because insects are sensitive to both visual and olfactory cues. Insects comply and pollination occurs. Perspective matters for how one views the interactions between flowering plants and insects. A plant-forward perspective may see insects as mere vehicles for transporting pollen, with flowers as clever lures. An insect-forward perspective may see pollen and nectar as mere food and pollination as just a byproduct of that attraction. The truth is somewhere in the middle. The plant is not doing something for the insect, it is enabling its own reproduction. Plants exploit a pollinator's mobility and its need for sugar—the same sweetness by which sugarcanes lure humans. The insect is not doing something for the plant, it is gaining nutrition for itself. Mutualisms are not mutual altruism, but mutual exploitation.

BEES

There are about 20,000 species of bees, classified in seven families, and found worldwide. Fossils of bees and bee nests have been found in rocks dating from about 100 million years ago, in both South Asia and southern South America. About 3,600 species of bees occur in the United States. Most people do not see many different bee species in everyday life and are unaware of the small, unassuming bees that make up the majority of species. Only a small number are familiar. Of the 3,600 bees in the United States, 49 are bumble bees—those slow-flying bees with large furry, black bodies with yellow patterns. They live in colonies of 40–60 workers and only one queen. There is one species of honey bee in the United States—western honey bees (*Apis mellifera*). The rest—3,550 species—are solitary bees, such as digger and leafcutter bees. "Solitary" means the bees do not live in colonies with a queen, which is the popular image of how all bees live. Nope. Solitary bees rule and they are important as pollinators and for their other contributions to the environment. Important they are, but not all bees are pollinators. For example, some bees are nectar parasites, taking nectar from flowers but failing to pollinate them.

Although all of these bees play key roles in the environment, our focus in this chapter is only one species.[4] You have seen them and where they live. If you drive in most rural areas of the United States, you will likely see stacked white boxes, set back from the road, usually at the edges of fields. Unassuming as these boxes may be, the animals living in them affect you every day, because of what you eat. Those white boxes contain western honey bees (Figure 13.1). Lots of them. What began in the wild as bees making food for themselves to survive times of hardship has been converted into a mutualism with us. Honey bees acquire safety from us, and in return, they make us honey and pollinate our crops.

All honey bees are organized into a single genus—*Apis*—that has been around for at least 60 million years, based on well-preserved specimens in amber. Believe it or not, honey bee taxonomy remains unresolved, because it is not clear how many species there are, with estimates ranging from six to eleven. Despite this uncertainty, some truths are accepted. Species are clustered into three groups. The first includes the dwarf honey bees, with two species from southern Asia/northeastern Africa that make a single open honeycomb on tree branches. The second group includes the giant

Figure 13.1 European (western) honey bee, *Apis mellifera*; note pollen basket on hind leg.
Source: Courtesy Matt Bertone

honey bees, with between two and four species from southeast Asia that construct a single open honeycomb under tree branches. The third group includes four species that build nests within cavities, such as hollowed trees. Three of those are restricted to southeastern Asia, with the final species, *Apis mellifera*, found worldwide, moved around by humans.

There is considerable variation within western honey bees (*Apis mellifera*), and over 30 subspecies have been described. For example, *Apis mellifera mellifera*—the subspecies called European dark honey bees (when three words are given for a scientific name instead of two, the third identifies the subspecies)—were originally only known from Europe, where they lived for the past 300,000 years in a climate with cool summers and cold winters. But this subspecies didn't stay in Europe. They were brought to the United States in 1622, and were the only honey bees on the continent until the mid-1800s when another subspecies was introduced (Italian honey bees, *A. m. ligustica*). Since then, spreading and mixing of various subspecies has occurred worldwide.

Honey bees are one of few animals that are "truly social" animals, better referred to as "eusocial." Being eusocial involves more than associating or living with others. There are degrees of sociality, but eusociality is defined by a set of characters: (1) common nest site; (2) cooperative brood care; (3) reproductive division of labor—not all individuals reproduce; (4) overlapping generations, with adult offspring interacting with a parent; and (5) sharing food with one another. Honey bees meet these criteria: they live together in a "hive"; worker bees take care of the young; one queen reproduces, the others take care of her and the hive; the generations overlap, so worker bees—which are all female—live long enough (about 4–6 weeks) to help care for their younger sisters; and foraging worker bees bring back food that they pass on to other workers in the hive, in a practice called trophallaxis.

Eusociality is maintained in part by pheromones, which are specific chemical compounds produced by both the queen and some workers. Pheromones emitted by the queen also regulate attraction of males to the queen. Outside the hive, males are receptive to the pheromone and will mate with the queen, whereas males are not receptive to the same chemical if it is detected inside the hive. Many Hymenoptera are eusocial species, such as bumble bees, hornets, and all ants. A few other insects have evolved eusociality, including wood-feeding cockroaches (eusocial cockroaches are better known as "termites"[5]), some gall-forming aphids (e.g., *Pemphigus spyrothecae*) and gall-forming thrips (e.g., *Kladothrips*), and even one ambrosia beetle from Australia (*Austroplatypus incompertus*). Outside of insects, eusociality is much less common, having only arisen in three species of sponge-dwelling snapping shrimp (*Synalpheus*) and two different mole-rats (*Fukomys damarensis* and *Heterocephalus glaber*). As one can surmise from this overview, despite many familiar species being eusocial, the habit is not common among animals. Honey bees aren't unique for being eusocial, but they're among a special minority. Let's appreciate them as such.

Honey bee colonies are essentially made of workers and the queen, which are all female. Males, called drones,[6] are produced deliberately and in small numbers. The queen lays drone eggs in preparation for the colony to "swarm," which is when the bees leave to find a site for a new queen. The sole function of a drone bee is to mate with the queen. Reproduction in the Hymenoptera is termed haplodiploid.[7] Females produce eggs: if an egg is fertilized, the offspring develops sexually and is female with two sets of chromosomes, so is diploid. If the egg is unfertilized, it still develops, but asexually, and the offspring is male with only one set of chromosomes, so is haploid. The queen can choose to fertilize or not fertilize a given egg, which explains her ability to "deliberately" lay drone eggs. Haplodiploidy is what enables eusociality in honey bees—though not in termites. Honey bee workers share 75% of their genes with their sisters, but they share only 50% of their genes with daughters. So, worker bees can help perpetuate more of their genes by taking care of their closely related sisters than if they reproduced themselves. In the honey bee colony's sharp division of labor, only the queen lays eggs. She can lay about 1,500 eggs per day, or 60–80,000 eggs per season, and she can live for several years.

What about defense? After all, we know honey bees, like most bees, sting (not all bees![8]). In fact, what is a sting?[9] In short, there are two components—a venom gland and delivery needle, a sting—and both have fascinating stories. The structures that make up the sting are actually remnants of legs of early arthropods that have been modified to lay eggs, called an ovipositor. Along with structures such as wings, the evolution of the ovipositor in wasps is among the most iconic stories in insects. Initially, the structure was blade-like and used to saw into plant tissue, with an egg laid in the wound—a strategy still used by sawflies. Some sawflies changed tactics, drilling into wood rather than softer plant tissue—a strategy still used by woodwasps. Woodwasp ovipositors are long, cylindrical, pointed, and used to drill into wood. This long pointy shape was co-opted for parasitism and used to lay eggs on or inside insect hosts—a strategy still used by parasitic wasps. Finally, the components of the ovipositor fused together to form a sharp needle delivering venom, the sting—a strategy used by the stinging wasps, including hornets, paper wasps, ants, and bees. Wasps with an ovipositor modified into a sting no longer lay eggs through it, instead eggs come out of a new opening at the base of the sting.

The honey bee ovipositor is modified into a sting to deliver venom. The sting is derived from the ovipositor, and only females have an ovipositor; therefore, only females can sting. Most stinging wasps use their sting for defense and for paralyzing prey used as food for their larvae. Being pollen-eaters, honey bees have no need to sting prey; their venom has a single specialized function, to induce pain in vertebrate nervous systems. This is true of most bees, but honey bees go a step further. Their sting is barbed. Once it lodges into the thick skin of a mammal, it doesn't come back out. In fact, the poison sac, which holds the venom, is pulled entirely out of the bee when it flies away. The sac is musculated, so it continues pumping venom on its own. But often some of the digestive tract and nervous system are also torn out of the bee's body. That trauma to the honey bee means she can sting only once, after which she dies. But that is for worker bees defending their hive. How many times can a drone sting? The alert reader already knows. It is a question that students miss on exams each year.

As vicious as honey bees sound, it may be surprising to learn they are actually not that dangerous to humans. True, honey bee stings can be fatal. According to the Centers for Disease Control, an average of 62 Americans died each year during 2000–2017 from stings of bees and wasps. As awful as each death is, 62 per year is a relatively small number. This average does not parse honey bees from various hornets, wasps, ants, and other bees, which would drop fatalities from honey bees much lower. Dogs kill more people than honey bees. Bee stings are generally not fatal on their own, but bee venom contains enzymes that can cause a systemic allergic reaction, called anaphylactic shock. Anaphylaxis is no joking matter. If you suspect sensitivity, take bee stings seriously! We caution against societal demonization of bee stings, but we urge individuals to manage personal risk appropriately.

BEE BEHAVIOR AND PRODUCTS

Each of those white boxes set back from the roadsides contains one queen and thousands of her female offspring, who are the nest's workers. Foraging worker bees bring back pollen and nectar to the hive to share with other workers who take care of their sisters and the queen. Foraging bees initially find flowers by flying in a random direction. Once a source of flowers and nutrients (or of water) is found, the bee returns to the nest and performs a "dance," as way to communicate complex information to the other foragers, conveying distance and direction to the resource.

Two kinds of bee dance occur: a "round dance" and a "waggle dance." The round dance tells other bees that a resource is found nearby, generally less than 100 meters from the nest. The waggle dance is more complex, involving movement in a straight (vertical) line as well as circular movement. The angle deviating from vertical approximates the angle between the food source and the sun and so conveys the direction to the food. The distance to and the quality of the food source are proportional to the time it takes for the bee to complete its wavy, waggle dance.[10] The interpretation of the bees' dance language by ethologist Karl von Frisch in his 1927 book, "The Dancing Bees," was met for years with skepticism and derision. Both derision and skepticism diminished greatly once Frisch was awarded the Nobel Prize in Physiology or Medicine in 1973.

Foraging bees convey information about nectar and pollen, both of which provide nutrition for the hive. Nectar provides carbohydrates, and pollen provides protein, vitamins, and lipids. Bees see color, with their visual reception shifted toward the shorter ultraviolet wavelengths: bees see purple and blue better than they see red. Then why are they attracted to red flowers? Red flowers may be perceived as red in the wavelengths that humans can see, but the flowers may reflect patterns visible in ultraviolet that bees can see but humans cannot (Figure 13.2). Seems like a lot is going on with flowers. Why? Easy. While the worker bees are collecting the pollen and nectar for their sisters, they are pollinating flowers. Pollen sticks to bees, which are extremely well adapted to deal with it. First, bees are the only insects whose "hairs" (setae) are branched, which helps hold pollen grains. Pollen grains carried by a bee then get transferred when the bees visit other flowers. At some point, the bees groom themselves with specialized pollen combs on their hind legs, and the pollen

Figure 13.2 Black-eyed Susan in visible (to humans) light (left) and in ultraviolet light (right); note the "nectar guides" that may lead honey bees to the flower's pollen source.
Source: © Prof. Andrew Davidhazy, www.davidhazy.org/andpph

is pushed into further specialized structures on the hind leg called pollen baskets (Figure 13.1). Honey bees can carry an estimated half of their body weight in pollen, slightly more in nectar. On each foraging trip, a honey bee collects either nectar or pollen, but not both on the same trip.

Honey bees undergo complete metamorphosis, meaning their life cycle includes eggs, larval stages, a pupal stage, then the adult. The time *A. mellifera* needs for development from egg to adult varies, differing among queens, workers, and drones, and also depending on temperature. Generally, development requires about 3–4 weeks at optimal temperature ranging from 33–36°C. That optimal temperature is regulated by worker bees fanning their wings inside the hive to keep it from getting overheated, or by crowding together to warm the hive and its developing young in their individual cells. Honey bees develop in hexagonal cells made of beeswax, produced by special glands on the underside of the adult bee's abdomen. Beeswax is produced by the metabolism of honey in the bee's fat cells. Beeswax is energetically costly to make—it takes the energy from about 8 kilograms of honey to yield 1 kilogram of beeswax.

OK, that's the hive, beeswax, pollen, and nectar. Where does honey fit in? Honey.[11] Ambrosia, amrita, food of the gods, drink of the gods. Lofty names and descriptors tell of humanity's love of the viscous fluid made from nectar. In a term somewhat less lofty, honey also has been called "bee vomit," though that term is not truly accurate. Returning foraging bees pass on the nectar from their "bee stomach" (an enlarged part of their foregut) by regurgitating it to other workers in the hive. Bee vomit, moth spit . . . oh, the world through an entomologist's eyes.

Nectar is predominantly water but contains sugars and other molecules that flavor it and it mixes with enzymes in the bee stomach. That mixture is regurgitated onto the walls of cells in the hive. When bees fan their wings, they are not just regulating temperature; the movement of air over the cells starts the process of dehydrating the liquid mixture, causing the water content to drop to about 18%, and making honey. It takes about 4 million flower visits to collect enough nectar to be converted to a kilogram of honey. Or, for the same kilogram of honey, bees must fly more than 160,000 kilometers to gather enough nectar. The term "busy as a bee" is not an empty phrase! The next time you eat a teaspoon of honey, realize that about a dozen honey bees foraged throughout their lifetime to produce it.

ORIGIN AND MOVEMENT

The European subspecies, *Apis mellifera mellifera*, fared very well when European settlers brought their bees to North America in 1622. Because of the climatic similarity of the New England colonies to the ancestral home of *A. m. mellifera* in Northern Europe, bees that escaped from the new American beekeepers' hives were able to survive, become naturalized, and revert to their original lifestyle of nesting in tree cavities. Subsequent intentional introductions of European honey bees included three subspecies, one each from Italy, Turkey, and the Caucasus Mountains of western Asia.

One other introduction was less intentional—at least the introduction into North America. A colony of the subspecies *A. m. scutellata* was introduced from Southern Africa to Brazil in 1956, with the intention of developing a bee that could better produce honey in a tropical climate. A year later, an accidental release resulted in swarms that crossed with European honey bees and established in the wild. Subsequent spread of these "Africanized" bees in the late 1980s carried them northward through Central America and Mexico, with brief forays into the Southern and Southwestern United States. In 1990, a swarm crossed the border into south Texas and established a permanent colony there. Africanized bees have been moving northward ever since, although, as of 2020, cold winters have limited their spread.

After Africanized honey bees established, or as managed bees bred with them, fear of Africanized honey bees developed to an extent well beyond their true negative impact. The venom of Africanized bees is no more potent than that of European honey bees, but they are more ready to become defensive and aggressive, will pursue transgressors for up to a kilometer, and an encounter is more likely to result in many more stings, even though individual bees can sting only once. It is that fear of many stings, even hundreds, that created the fear of "killer bees" and spawned a mini-industry in Hollywood, as the bees were demonized and sensationalized on the silver screen.

Let's return to those white roadside boxes, whose residents we labeled earlier as "European." Did the bees evolve in Europe? Actually, no. The name "European" is accurate for the subspecies, but not for the species. At one time, *Apis mellifera* had been considered to have originated in Africa, but newer analyses have determined the ancestral home was more likely Asian, which is also the center of origin of the other *Apis* species.

BEEKEEPING BEGINS

All of that is well and good, but how did honey bees affect human history? For that, we need to consider beekeeping. Beekeeping? What about it? Humans have had a long relationship with honey bees, as have other primates, including chimpanzees and gorillas—all of whom collected honey and included it in their diets. No one knows when honey hunting by humans began, but it may date back to our origin as a species. Honey hunters were depicted at least 8,000 years ago on cave paintings located in present-day Spain. Honey hunting is still practiced by the indigenous Hazda people in tropical East Africa, who obtain 15% of their calories from honey gleaned from wild bee nests, especially during those times of the year when game is scarce—honey provides them with crucial calories. Some anthropologists and biologists believe that honey hunting may have been an integral activity in history, by providing the calories needed for development of larger brains in early humans, supporting the idea of ancient hunting for honey.

Honey hunting is also practiced today by people interested in bees. The activity requires finding a bee colony, being able to reach it, being able to evade or tolerate the stings of defensive bees, and having cooperators to help bring back the prize that the honey represents. Although finding a honey tree—a tree that contains a nest of honey bees—is not rocket science, it requires a systematic approach, a keen eye, and patience. Entomologist and bee expert Tom Seeley has spent decades finding natural bee nests. Most nests he has found have been 6–10 meters up in trees, with the opening into the tree a hole only a few centimeters in diameter. His searching in temperate North America has found that nests occur at low density, generally about 1–3 nests per square kilometer. The density may vary with different vegetation or geographic locale, but his results give some idea of natural occurrence of bee nests. That range of numbers actually matters; we'll return to it later.

Honey hunting for natural nests is not the same as beekeeping, which is providing a place for bees to safely build a nest, raise young, and produce honey and beeswax. It was only a small step to go from honey hunting, taking honey from naturally occurring nests, to tree beekeeping. Tree beekeeping involves a beekeeper chopping a long slot in a tree, then excavating the tree cavity and fitting it with a small door that can be opened to remove the bee products. Bee swarms naturally find the created cavities and build nests there. Because these bee "keepers" do not move or manipulate the bees directly—only harvesting part of the honey and beeswax that the bees produce, tree beekeeping has only minimal impact on the bees.

The activity we generally consider beekeeping began thousands of years ago. It is difficult to pin down exactly where and when, and it likely occurred independently in multiple places. Containers that were left outside—accidentally or intentionally—were colonized by bees. After humans found that bees constructed a nest in the container, the people then placed more containers to be colonized, which led, eventually, to making containers specific for the bees to colonize. Providing a place for bees to nest allowed for control, to some extent, of the harvest of the bees' products—honey and beeswax. Caring for bees by providing a home and protecting them enabled long-term relationships between beekeepers and their bees. Early beekeepers who provided a home used a "skep" for the hive, to be colonized by swarming bees. Skeps are usually dome shaped, as an inverted basket might be, often made of straw, and are still used by some beekeepers.

ANCIENT BEEKEEPING

The actual origin of "modern" hive beekeeping is not known, but hive beekeeping[12] occurred at least 4,500 years ago in Egypt and is documented in carved relief showing beekeeping activities. At that time, the hives were horizontal, made of clay, and sealed at one end. Rows of clay hives were stacked atop each other in a method that is still used by some beekeepers in the Eastern and Southern Mediterranean. Honey bees were an important part of Egyptian culture. Egyptian legend stated that the sun god, Re, created the world, and when he cried, his tears became honey bees as they touched the ground. Bees were well represented in the hieroglyphs that made up the written language of the Egyptians. The hundreds of Egyptian hieroglyphs included specific symbols to denote honey bees, beekeepers, and honey.

Honey had a number of uses in ancient Egypt. It was a sweetener of food—sugar and sugarcane had not yet made their way westward to Egypt—and was used in lieu of currency to pay debts or taxes. One can imagine honey would not have willingly been carried as "pocket change." One important use of honey was for medical purposes. Early Egyptian medicine, which dates at least to 2000 BCE, was considered by many to be the best in the known world—the world excluding Asia, anyway. By about 2000 BCE, some 900 medical treatments were known, of which about 500 included honey in some way, even if only to make the primary medicines easier to take. Some of the treatments also mention beeswax.

In one unusual medical example, honey was part of a treatment used as a contraceptive, in a concoction that included sour milk, sodium carbonate, honey, and crocodile dung. This treatment would seem to generate a number of questions. First, how much crocodile dung? Who collected it? Was "crocodile dung collector" a legitimate occupation? The collectors likely could not get life insurance. Second, who thought of crocodile dung, and why that specific kind of dung? Was its efficacy proven? Was there a safer alternative? Of course. Nearly *any* alternative would have been safer. Third, though we are not actually posing this as a question, we don't even want to consider how this recommended treatment was used. And finally, did anyone actually use this prescription for contraception? Perhaps anthropologists should examine ancient Egyptian population numbers and look for sharp increases or decreases to infer whether the treatment was used or ignored. Crocodile dung! And we thought toad vomit (Chapter 5) was more than a little odd.

Our friend Gene Kritsky, entomologist and Egyptian expert, wrote about the contraceptive example in his book, *The Tears of Re*. Among the many details of early Egyptian culture related in that book is the paradox that honey—the component of contraceptives—was also both considered an aphrodisiac and a treatment to increase fertility. The male god Min was the god of fertility and sexual potency. Some priests of Min were the beekeepers for temples, whereas others collected honey from wild nests. Another detail Kritsky related was the role—captured in some of the ancient hieroglyphs—of "bee queen caller": certain beekeepers could make the sound of a soon-to-emerge new honey bee queen, and they could determine whether the colony would soon swarm. This practice, though not without risk of being stung on the lip, certainly seems safer than being a crocodile dung collector.

Beekeeping was practiced in places other than Egypt. Archaeological excavations at Tel Rehov (in modern-day Israel), that date to about 1000 BCE, have uncovered more than 200 bee hives, including 30 intact hives, and even the remains of bees. Amazingly, the bee remains yielded DNA that allowed researchers to identify the subspecies of bee—*Apis mellifera anatoliaca*—which was most likely moved from its home in Turkey to the site at Tel Rehov. The configuration of hives at the site, in rows of 100, provides evidence of a significant beekeeping activity and industry that yielded honey and beeswax 3,000 years ago.

Over time, beekeeping gradually moved from the Eastern Mediterranean further to the east. Beekeeping went to Persia and, sometime after the beginning of the Achaemenid (Persian) Empire about 550 BCE, trade along the Silk Roads took the practice of keeping western honey bees to China. Beekeeping had been practiced in ancient times in China, but the Chinese historically kept Asiatic honey bees, *Apis cerana*, which had much smaller colonies. Chinese beekeepers were skilled in their craft, but the primary product they sought was beeswax. Beeswax was in great demand for candle making and was used as offerings to the emperor and for various other purposes. An interesting aspect of *Apis cerana* biology is its unique defense against one of its native predators, the Asian giant hornet, *Vespa mandarinia*.[13] When a hive of *A. cerana* is invaded by the hornet, hive worker bees surround it and flex their wing muscles, raising the temperature in the hive to about 46°C—hot enough to kill the hornet, but just below the honey bees' thermal limit of 48°C.

Just as the Chinese were careful observers of silkworms, so were Chinese beekeepers observant of *A. cerana*. But observations can be colored by preconceived notions or experiences in everyday life. The Chinese beekeepers, skilled at observation, were totally wrong about one significant aspect of their bees. They likened a bee colony to a royal family, living in a royal palace, led by a king, and served by working males who provided food and protection. They identified drones and correctly stated that the drones did not have venom, but they also thought that larger hive workers were drones, as well. Chinese beekeepers correctly identified the tasks of different bees, but they had the sexes wrong. Other than the drone, all the colony members they thought were males were and are, in fact, females. Drones do not have venom, but that is because they are males and, as mentioned earlier, male bees do not have stings. No doubt the associations were made based on the human societal gender roles of the time—the rulers were kings (or emperors), not queens, and males worked outside the home and were the homes' protectors. The beekeepers were very precise in their observations. But precision and accuracy are not the same thing, and the beekeepers were flat out wrong about the role of the sexes.

Mistaken gender roles or not, the Chinese were skilled beekeepers. When western honey bees arrived with Silk Road traders, the Chinese were able to transfer their knowledge and husbandry, and they readily adopted the newly arrived bees. Western honey bees produced more honey and allowed the Chinese to increase the scale of honey and beeswax production, even while maintaining colonies of the native species, *Apis cerana*.

Beekeeping also made its way to Europe, first moving along the eastern shores of the Mediterranean. After Alexander the Great conquered Achaemenid Persia in 333 BCE, the newly conquered territory became part of the Greek Empire. Apparently, this was the time when beekeeping went to Greece. Alexander's tutor, Aristotle, wrote about bees and the practice of beekeeping. As honey bees were moved to Italy, Roman writers picked up the job of chronicling their use. Bees eventually spread throughout the continent, and the role of primary beekeepers passed to the monks and nuns in the many European abbeys and monasteries. The hives provided wax for candles and the honey was used both for food and for the fermented drink called mead.

FLOAT LIKE A BUTTERFLY, STING LIKE A BEE

The saying "float like a butterfly, sting like a bee" was made famous by the renowned boxer Muhammad Ali. It was in 1964, while Ali was still known by the name Cassius Clay and the brash 22-year-old was preparing to fight world champion Sonny Liston that he made the statement during an interview about his strategy. Everyone knew that Liston would batter and beat Clay, but Clay danced lightly around the ring, staying out of reach until he got an opportunity to deliver his own fierce punches and won by technical knockout. Ali's (Clay's) fame skyrocketed and continued unabated long after he retired from boxing. Still living on, in addition to his humanitarian and anti-war efforts, is his quote, equating his lightning-fast blows to the sting of a bee. Unless Ali (Clay) was referring to other species of bees, he overlooked the fact that honey bees can sting only once. Or maybe he thought that just one of his solid punches was all that was needed to vanquish his opponent. There is no record of Ali ever studying the biology of bees, so we will cut him some slack.

Recognition of the offensive power of honey bees was known long before the 1964 Liston-Clay fight. In fact, the use of bees as weapons went back at least as far as the 2nd Century BCE. In 113 BCE, Mithridates VI, self-proclaimed descendant of Alexander the Great, declared himself ruler of the Kingdom of Pontus, which encircled the Black Sea and occupied modern-day Turkey, Armenia, and southern Russia. A series of three Mithridatic Wars with the Romans occurred during 88–63 BCE. Bees made their appearance and earned their fame during the Third Mithridatic War. The Romans had laid siege to the city of Eupatoria, where the army under Mithridates had holed up. Roman troops dug tunnels under the city walls with the intent of emerging behind the walls and capturing the city. According to Jeffrey Lockwood, in his book *Six-legged Soldiers*,[14] Mithridates and his troops released bees into the tunnels (they also used other deadly tactics), thus breaking the siege and anchoring the historical role of honey bees as weapons. Later, as Mithridates was retreating across Anatolia (modern-day southern Turkey), he had his troops strategically place containers of honey along the route. The act was a ruse, and the intent was that the honey would be found by the Roman troops, as if the containers had been discarded or left during the retreat of Pontian troops. However, the honey was poisonous, having been produced by bees feeding on rhododendrons. The Roman troops duly found the "discarded" honey and consumed it; they were sickened by the toxins it contained, incapacitated, and then killed by supporters of Mithridates.

The use of bees as weapons continued through the next two millennia. Castles were built with niches, called bee boles, in the stones of south-facing interior walls, which held bee skeps. South-facing walls gathered sunlight, warming the bees so they were more likely to be active. If the castles were attacked, defenders dropped hives of angry bees onto the invading forces. Honey bees even figured in the Crusades. When Richard I left England to wage war against Muslim rulers in Jerusalem, his 200 ships included 13 holding bee hives. The bees were used in the Siege of Acre in 1191, when bee hives were catapulted by trebuchets over the city walls to bombard the residents and troops within. Curiously, the word "bombard" has its origin in the

Greek word "bombos," Latin "bombus," for a buzzing or humming sound, and *Bombus* is the name of the genus of bumble bees (not honey bees). One can imagine that a colony of bees—whether *Apis* honey bees or *Bombus* bumble bees—would not have been happy about being placed into a catapult or trebuchet; the bees were even far less likely to be congenial upon their landing.

Keep the same technology but jump ahead by a thousand years, even though bees were used as weapons in all sorts of wars in the intervening period, to World War I, when agitated honey bees helped in the defeat of the British Expeditionary Force in the 1914 Battle of Tanga during their offensive in German East Africa. The British assault was not going well, and the attacking bees further demoralized the infantry and spurred their retreat; the defeat was described as one of the worst in the history of the British army. What about Gallipoli? In British propaganda, stories were told of the German defenders deploying bees and trip wires that caused hives of bees to drop onto the advancing troops. In reality, the bees also attacked the German defenders, being non-partisan stinging weapons.

More than fifty years later, bees and trip wires *were* deployed as booby-traps in South Vietnam by Viet Cong forces. Different accounts offer varying versions. In one account, the Viet Cong attached small explosives (e.g., firecrackers) to the huge aerial nests of *Apis dorsata*, causing them to drop and releasing angry bees onto the advancing American troops. In another version, trip wires resulted in *A. dorsata* nests dropping onto American tanks, with angry bees entering into the close, confined space. *Apis dorsata* is the largest and most aggressive of the Asian honey bees; the tank's crew had the choice of staying in the tank and facing many large, aggressive bees in tight quarters or trying to exit the tank quickly, only to find themselves under fire from Viet Cong—not very good choices. Trip wires releasing bees were also used as booby traps in the maze of tunnels used by the Viet Cong under their villages; again, being trapped in a narrow tunnel with angry bees would not have ended well.

EXPANSION TO THE WEST

OK, that is a lot of history. We traveled from Egyptian beekeepers and medical prescriptions to weaponized giant honey bees in the 1970s. That still doesn't explain how honey bees impact you. To learn about that, we need to go back again. Again? Bear with us. In this case, we go back 400 years, to one more movement of honey bees, in this case along a different route, but still a route of the extended Silk Roads.

Previously we mentioned the movement of honey bees from Europe to the Americas. The cross-Atlantic movement of honey bees in 1622 was made in peace, not to initiate or aggravate war. European emigrants are considered to have brought western honey bees with them as they moved to their new homes in North America. By some accounts, the Native Americans called honey bees "white man's flies." The honey bees were clearly distinguishable from native bees, and sighting honey bees foretold the encroachment of white settlers onto indigenous people's lands. Why did the colonists bring honey bees from Europe?

Where do you think many of the foods we eat come from? In what has been called the Columbian Exchange, plants, animals, and diseases native to Europe and Asia were transported to the Americas, and plants from the Americas were transported to Europe and Asia. OK, and so? Many of the foods consumed in North America are products of plants whose ancestral home was Europe and Asia, and to a lesser extent Africa and the Pacific Islands. Transported food plants native to Europe and Asia and moved across the Atlantic include numerous fruits, such as apples, peaches, pears, grapes, plums, and citrus species. Vegetables that are native to Eurasia and were brought to North America include carrots, onions, celery, and broccoli (as well as other *Brassica* crops). In the food plants' native homes in Eurasia, they were pollinated naturally by honey bees—the bees lived there, after all. But they were not pollinated—at least not pollinated well—in their new North American home. Hence, the importation. Some plants that are native to the Americas, such as blueberries and squash (e.g., melons, pumpkin) can benefit from pollination by European honey bees, even though they have their own native pollinators. Pollinating crops is big business. Just how big, we will see shortly. But first, let's consider honey.

Certainly, you are aware that honey is an important agricultural commodity. According to the National Honey Board, a study from the University of California found that Americans in 2017 consumed 271 million kg (596 million pounds) of honey, nearly a kilogram per person. That amount had increased by 65% from 2009 to 2017. The same study showed that the honey industry contributed $2.1 billion (USD) to the 2017 US Gross Domestic Product, including the wages for 22,000 jobs. Beekeepers in the United States produced just over 70 million kg of honey in 2.8 million hives, meaning that a large amount of honey was imported from elsewhere. Worldwide, the honey market was valued at $7 billion in 2016, for honey alone, not including the economic additions and multipliers. According to Statista, in 2016, 1.78 million metric tons was produced around the world in 90 million hives. China produced the most honey—nearly 450,000 metric tons—but India led the world with the number of hives, at 13 million.

Honey is an important commodity and is big business, indeed. But honey doesn't hold a candle (beeswax, of course) to the value of pollination. According to the Pesticide Action Network, 100 crops produce 90% of the world's foods and, of those 100 crops, 70 are pollinated by bees, mostly honey bees: in North America, one in every three bites of food eaten would disappear if honey bees vanished. At least 95 kinds of fruit crops in North America are pollinated by honey bees. Worldwide, 10% of the total economic value of agricultural production for human consumption is derived from insect pollination, or about $200 billion (USD) in 2005. In the United States alone, commercial beekeeping contributes $20 billion (USD) every year to the country's economy, 10 times more than the contribution from honey.

IMPORTANCE OF BEES TO ALMONDS

Let's tease apart the pollination enterprise and focus on just one crop—almonds. The amount of land in the United States dedicated to growing almonds has increased steadily and, as of 2020, stands at 600,000 hectares, or 6,000 square kilometers. In English units, that is 2,316 square miles, or nearly the size of the State of Delaware. Maybe that does not seem like a lot, but that is land devoted to one crop, and that crop is totally reliant on honey bee pollination. Without bees, no almonds would be produced. The economic value of almonds in 2018 was $5.5 billion (USD); it is the third-most valuable crop in California and represents 11% of California's agricultural output.

In 2020, almond production was estimated at 1.4 billion kilograms, up nearly 18% from the year before. To achieve that yield, flowering almond trees need to be pollinated, and that is what honey bees do. There are nowhere near enough honey bees in California to pollinate the 600,000 hectares of orchards. As a result, honey bee hives are rented at an average price of $200 per hive (Figure 13.3) for roughly two weeks, shipped in like six-legged livestock by the semi-trailer load from all over the country. During the few weeks that almonds are in flower, more than half of all the managed honey bee colonies in the United States are in California almond orchards.

Read that again. More than half of all managed honey bee colonies in the United States are working in California almond orchards for two weeks, and they are responsible for producing 11% of the total agricultural output of California. So, something like 1.4 million hives rented for an average of $200? Plus, in the United States each hive is worth—not the rental, but its value—upwards of $300. That is about $400 million of resource sitting outdoors, in the open, often in remote areas. Each year, that resource attracts the attention of "bee rustlers" who arrive in the dead of night, load large numbers of the bee hives onto their own trucks, and disappear. In 2016, more than 1,600 hives were stolen from almond orchards during the flowering

Figure 13.3 Bee hives for pollination in an almond orchard, Central Valley, California.
Source: Richard Thornton/Shutterstock.com

season between January 1 and February 28. Some bee managers or orchard owners have been forced to hire armed guards to prevent rustlers from making off with this valuable commodity.

It seems that the honey bees are a pretty valuable resource, one that should be cared for and protected.[15] And it seems that the honey bees could be cared for and protected better than they are. We are not referring to rustlers. But, today, honey bees are in trouble, because of how we treat them. Tom Seeley takes the position that humans have not changed the genetics of honey bees as we have other livestock but, instead, we have changed the bees' environment. We manipulate the honey bees' habitat by moving them to where their pollination services are needed or where the bees will forage on plants to make a certain, desired honey crop. In both cases, the bees are crowded together and treated like livestock. Here is where Seeley's earlier estimate of the density of natural bee hives matters. His estimate was 1–3 per square kilometer. For grins (and easy math), let's say 2.3 per square kilometer. That means that the density of honey bees pollinating almonds in California orchards is 100 times greater than the natural density in the wild. Now, admittedly, the comparison is not truly fair, because the state's almond orchards are filled with an estimated 3.3 trillion flowers, and the bees are present for only a few weeks. But even with all those almond flowers available, the "farmed" bees are still crowded together, and crowding raises the risk of disease and infection.

BEES IN TROUBLE

We hope not to seem critical of beekeepers, but stresses on bees are huge. Bees shipped to California almond orchards are, after almond pollination ends, shipped to a next stop for another crop, and on again. Some bees go to pollinate apples, some cherries, others other crops. Many bees are "on the road" for months. Honey bees are under unprecedented stress, from being on the road, crowded together, and exposed to diseases and parasites. Whether as traveling bees or as residents of roadside boxes, bees are exposed to insecticides, and bees are especially vulnerable because of their biology. Some studies show that accumulation of certain classes of insecticides can either disorient or kill the bees. Controversy over certain insecticides pits growers against other growers, growers against government agencies, and even researchers against other researchers.

All the while the stresses continue and multiply and include: parasites, such as Varroa and tracheal mites; bacterial and viral diseases that target stressed bees; land cover changes that eliminate the flowers that provide pollen and nectar; and insecticides, which in some cases, are the final stress. Honey bees disappear in what is called colony collapse disorder, with no single blame. Death by a thousand cuts. Bee numbers are declining, in some cases precipitously. There are solutions, but they require thinking differently. Jonathan Lundgren, founder and director of the Ecdysis Foundation and Blue Dasher Farm, an independent research farm and laboratory in South Dakota, focuses his research on regenerative agriculture to enhance bee health.[16] Lundgren knows what many of us ignore: threats to bees represent threats to everyone. He has said, "Starve them [bees], take away their home, stress them by shipping across the country, infect them with diseases and parasites, then we poison them—is it any wonder they are in trouble?"

From their ancient origins, honey bees have influenced and shaped multiple cultures and empires. For centuries, honey bees were a central foundation of Egyptian cultures. Bees were transported along the Silk Roads, from the eastern Mediterranean to Central Asia and further east to China. Honey bees figured prominently in war and in peace. They were used as very effective weapons, at the junction of technology and biology. The world's food supply relies to an enormous extent on honey bee pollination. Agriculture—at the very least North American agriculture—developed over the past four centuries in concert with pollination provided by honey bees, which allow humans to create and consume a diverse food supply. We rely on *Apis mellifera* for production of fruits and vegetables. Take away honey bee pollination and then imagine going to the grocery—many favorite foods will be missing, for no replacement for honey bee pollination is available. Unless, perhaps, the modern, mechanized, misnamed "drones" could be repurposed for pollination. Honey bees have survived for millennia and, just in the last few thousand years, have impacted humanity. Here's hoping they can continue to provide that impact for millennia to come.

NOTES

1. It is not uncommon for other species to aid—or be critical to—another species' reproduction. For example, yeast-like *Adlerocystis* are associated with soft ticks (Argasidae) and are thought to help survival of sperm cells.
2. This condition is called anisogamy and is familiar to us (sperm and egg). But not all sexually reproductive organisms have males and females (i.e., smaller and larger gametes); some organisms have gametes that are the same size and fuse together during mating—a condition called isogamy, found in some algae and fungi.
3. One unusual example of wind-pollinated plants is *Welwitschia mirabilis*, an uncommon, relic species of an old lineage, found in the coastal deserts of southwest Africa. In 2008, RNW was with a small group of birders in western Namibia. On the itinerary was a visit to the "*Welwitschia* forest." Several of us had never heard of the plant; a few had. We anticipated a forest of ancient plants, hopefully holding some unique birds. The guides drove us for more than an hour over washboard roads through some of the driest desert on Earth—the Namib gets its only moisture from fog that rolls in from the Southeast Atlantic on the cold Benguela current that brings no rain. There was one stop, at an overlook with a view across deep, rough canyons, prompting colleague Michael Jeffords to describe the scenery as "like the Badlands, only badder." On we went. Finally, we stopped. Desert as far as we could see in all directions. We walked for a few minutes and there in front of us was a (as in *one*) prostrate plant, extending 3–4 meters across. One. That was the "forest." No unique birds. Who knows how far away the nearest neighboring *Welwitschia mirabilis* individual was located. Seeing the isolated plant provided a fresh insight into wind pollination—also how tenuous the existence of that plant was.
4. There are more books written about honey bees than about any other insect. Next to butterflies, honey bees are the best-known and most recognized insect, and equally the most misunderstood insect around. We do not attempt to compete with all those

books—we focus this one chapter on a little history and how bees impacted—and continue to impact—humans. We do not provide the definitive word on bees. There— we said it. We are not bee experts. Honey bees and beekeeping are mostly inseparable. Two excellent, but very different, books about honey bees are from entomologists Gene Kritsky and Thomas Seeley. For a rich exposition of the history of beekeeping with details of honey bees and their history, see G. Kritsky, *The Quest for the Perfect Hive: A History of Innovation in Bee Culture* (New York: Oxford University Press, 2010); also, *The Tears of Re: Beekeeping in Ancient Egypt.* (New York: Oxford University Press, 2015). T. D. Seeley, *The Lives of Bees.* (Princeton, NJ: Princeton University Press, 2019), tells how the honey bee that we benefit from in everyday life functions in the wild, not as a semi-domesticated species. Countless other books are available to interested readers.

5. Termites are social, but are diplodiploid, so they are the exceptions to the definition.

6. The modern flying devices used for surveillance, photographs and as weapons are called "drones." Given the extensive flight of honey bee workers, perhaps the appropriate name for the flying devices would be "workers."

7. The actual term is arrhenotokous parthenogenesis. In this type of reproduction, females develop by sexual reproduction (fertilization of an egg by a male) whereas males develop asexually, with no fertilization. Consequently, the male has only one set of chromosomes from his mother, so is haploid.

8. Stingless bees are another subject entirely. The 500+ species have fascinating biological stories and produce a honey that is 10-fold more costly than that from honey bee.

9. "Sting" is both a verb and a noun that describes the structure (rather than "stinger").

10. Details about recent interpretations of the fascinating nuances of the waggle dance are told beautifully in Seeley, *The Lives of Bees*.

11. Listing sources of information about honey is like listing information on bees—there is far too much to try to cover. Our intent for the presentation in this chapter is to provide just enough to answer questions raised by the biology of honey bees. As anyone who loves honey will attest, the richness and variety of complex flavors of honey rival the variety of taste and aroma of coffee. For those who have only ever had grocery store clover honey, seek out honey from other sources. But, as we relate in later paragraphs, steer clear of honey made from flowers of rhododendron or azalea!

12. For a delightful, detailed illustration of beekeeping in the early days of Egyptian cultures, see, Kritsky, *The Tears of Re*. Many of our details about Egypt come from examples in his book.

13. The Asian giant hornet, *Vespa mandarinia*, was first reported in the Pacific Northwest of the United States and Canada in 2020. Nicknamed "murder hornets," they may harm bee hives but are not considered any more dangerous to humans than honey bees or other wasps.

14. J. A. Lockwood, *Six-legged Soldiers: Using Insects as Weapons of War*, (New York: Oxford University Press, 2008) is the definitive word on insects as weapons. Jeff's penchant for arcane detail and excellent story telling make for a book that reads like a fast-paced novel.

15. In fairness, we need to state that western honey bees are also associated with declines of native solitary bees, and some native bee (and native plant) enthusiasts suggest that honey bees harm the environment—much as other livestock can cause environmental harm.

16. Jon Lundgren is a remarkable individual. Full disclosure: Jon was a PhD student who studied under RNW at the University of Illinois. For more than a decade, Jon was an award-winning author and cutting-edge scientist with the US Department of

Agriculture. At the peak of his career, he left his position because he believed that agricultural science was not addressing the root causes of or long-term solutions to crop and bee health. In 2016, he started the independent research foundation Ecdysis, which is conducting research to develop strategies for regenerative agriculture. In parallel with Ecdysis, he began Blue Dasher Farm, recognizing that, to be credible to farmers, he needed to practice what he preached. Blue Dasher Farm grows crops and animals to demonstrate and share regenerative approaches with growers from around the world. More information about their work is at: https://www.ecdysis.bio/.

SECTION 6

Tying the Silken Threads

CHAPTER 14

⌁

Tying the Silken Threads

Five insects. Just five. Who would have thought they could have been so significant or that they could be linked together by the metaphorical silken thread? We
introduced five insects that have had an impact on history, both positive and negative, ranging from small scale to world changing. How did we consider the impacts of
these five? We tend to value what we can measure, knowing full well that it may be
the unmeasurable things insects have done that matter the most.

IMPACTS—MEASURABLE AND IMMEASURABLE

Insects have made positive contributions to human history. Honey bees pollinate
600,000 hectares of almond orchards in California, providing added value to the tune
of $5 billion for that one crop. The overall value of honey bees as pollinators in the
United States is estimated at $20 billion, with the value of the honey they produce
estimated at $2 billion. The monetary value of honey bees can be measured, but the
value of the food they make available to us is not so easily quantified.

The dogged determination of Carlos Finlay, Walter Reed, and Jesse Lazear led to
success in the war against yellow fever in Cuba, and the plan by William Crawford
Gorgas enabled the completion of the Panama Canal. The economic benefit of the
canal was (and is) huge—more efficient shipping, reduced distances traveled, and
safer transport. Defeating yellow fever mosquitos and yellow fever made that possible. Would the canal have been built if mosquitos and disease had not been controlled? Perhaps. But by one estimate, implementation of Gorgas's plan saved the
lives of around 70,000 people who would have been needed to work on the canal had
yellow fever not been controlled. The economic value of the canal can be counted, but
cannot be compared to the lives saved.

The global value of silk is expected to reach $17 billion by 2021. But can we mea-
sure the value of the silkworm, whose product not only directly spawned innovations

The Silken Thread. Robert N. Wiedenmann and J. Ray Fisher, Oxford University Press. © Oxford University Press 2021.
DOI: 10.1093/oso/9780197555583.003.0014

but whose trade connected cultures, ideas, religions, and people? The Silk Roads of ancient times are not just part of a historical tale—the routes are vibrant today and still expanding. The New Silk Road, a construction project initiated by the Chinese, forms a network of rail and sea routes that mimics the original trade routes, though the camel caravans of old have been replaced by "iron camels." The New Silk Road, also called the Belt and Road Initiative, the "Belt" being land routes and "Road" being sea routes, reaches 45 countries. The impact of the Silk Roads, resulting from the silkworm, is immeasurable.

Not all impacts of insects were positive. Oriental rat fleas, as the reservoir for plague bacteria, initiated three plague pandemics that killed hundreds of millions of humans. Lice were a near-constant part of human life for millennia. Long known as the vector of a horrible disease—typhus—we now know they extended the plague pandemics initiated by the rat fleas. Yellow fever mosquitos—the name is an anomaly. Today, *Aedes aegypti* has been replaced by other mosquitos as the most important vector of yellow fever in South America. Also, its name belies its importance, as yellow fever mosquito is the vector of other serious diseases—for example, as many as 400 million cases of dengue occur annually, and that number is increasing. And while yellow fever mosquitos did not cause slavery, they certainly contributed to the industrial scale of the slave trade. The deaths of slaves from yellow fever made repeated shipments of replacements necessary and thus contributed to the devaluation of the lives of slaves.

HISTORY THAT IS NOT FOUND IN HISTORY BOOKS

We have also endeavored to tell some of the stories behind the statistics and impacts, the parts of history that none of us likely learned early in life. As we stated from the outset, we are not professional historians. Insect-related stories behind conventional history ought to help illustrate why and how some historical events unfolded. But for some insects, if their history is not known, their stories are all but impossible to tell.

Back stories abound for the insects we included in this book: stories about the fleas and plague that yielded the concept of quarantine that we still use 600 years later and about Alexandre Yersin's discovery of plague bacteria in a grass hut, not in a high-powered research laboratory. Buried below the surface of the tales of silk and the silkworm are the stories of the women's revolt against the silk ban; of Louis Pasteur and Agostino Bassi saving the silk industry; of the worker's revolt over the introduction of the Jacquard loom; and of the quashing of that revolt—an oft-repeated modern response. We wove the threads from the Jacquard loom to modern cell phones, each additional thread adding to the story, with the complete woven product real, not just part of a tale. Yellow fever mosquitos were connected to slavery, certainly, but they also played a part in the legacy of Theodore Roosevelt. The disease carried by those mosquitos led to a bizarre plot to kill President Lincoln with clothes soiled by yellow fever victims. The story of yellow fever in the Americas only came about because the thread running through a series of events remained intact over the years, running from the Silk Roads all the way to the Yucatan Peninsula of Mexico. If any one—*any one*—of the series of events on that thread had not occurred, the unlikely possibility of yellow fever reaching the Americas, a rare event, would not have been realized. We have not seen this entire thread, or an explanation of how unlikely the arrival of yellow fever in the Americas was, in other histories of slavery. In this book we also told of bees that were ancient weapons, but the dropping of *Apis dorsata* hives onto military tanks in Vietnam reflected a 2,000-year-old difference in weaponry between two armies. These back stories failed to make it into history books, at least the ones we have encountered.

Some of the insect stories told of inhumanity, such as the way victims of insect-borne disease were treated before anyone understood those diseases or the forcing of Polish Jews into an overcrowded ghetto so that lice and typhus could do the work of the Nazis. Still others told of the horrific use of insects as vectors of disease in biological warfare. But other stories also told of humanity, such as the aid provided to typhus victims by the Grey Nuns of Canada, the volunteers for yellow fever experiments in Cuba, or the often-sacrificial assistance of doctors, nuns, priests, and prostitutes in Memphis in comforting the thousands dying from the scourge of yellow fever. A few stories were just fun, such as the tangent on pubic lice, and a few were head-scratching (not from head lice), such as Newton and his toad vomit, or the ingredients mixed with honey to serve as a prophylactic in early Egypt.

We did not tell all the possible stories of insects impacting humans. How do we measure the impact of the mosquitos that transmit malaria? The 400,000 people that die each year, every year? We can count deaths, but we can't quantify lives. Mosquitos that transmitted malaria also affected the outcome of fighting in the Pacific Theater in World War II. Because the US military had DDT—while the Japanese did not—fewer Allied troops were debilitated by the disease. We did not tell of the insects that cleared vast tracts of Australian grazing lands of exotic cactus, returning worthless land to productivity. We did not tell about the beetle and tiny parasitic wasps that saved the citrus industry worldwide—orange juice is available as a breakfast drink in the United States because of a few beneficial insects. We did not include the weevil that nearly ruined the US cotton industry and caused a great migration of people in the United States. Nor did we include stories of desert locusts, whose huge swarms have stripped vegetation from crop plants and caused food emergencies in East Africa in the past and still today.

The impacts of some insects don't fit any simple measure. Fruit flies, for example, those tiny flies that seem to suddenly appear when bananas get a little too ripe, would seem to be a pest that we would want to be rid of. However, as a research organism, fruit flies underpin our understanding of genetics. Thomas Hunt Morgan won a Nobel Prize in 1933 for his work on fruit fly genetics, demonstrating that genes were carried on chromosomes. He was influenced by the work of Nettie Stevens, who figured out the role of x-y sex chromosomes by studying darkling beetles, which may be pests in your pantry right now. The understanding of genetics is improving human health today, and the future roles of genetics for disease mitigation are beyond calculation. The many as yet untold insect tales would fill another book. Hmm . . . there's a thought.

Never doubt for a moment that insects have affected human history. To empha-size that, we conclude by telling the story of one significant figure in history. All five insects we included in our stories intersected with this one person, three of them in a big way. Napoleon Bonaparte has the distinction of being the only military leader (or at least the only one we know of) who was defeated by insects three times. Defeat of a mighty general by insects would be almost comical if the impacts on history of all three defeats had not been so enormous. Let us finish our narrative with these stories, connected, as it were, by the silken thread.

EGYPT, SYRIA, AND FLEAS

The Napoleonic incursion into Egypt and Syria should never have happened. The inva-sion was solely the empire-building continuation of recent victories by Napoleon's army in Italy against Austrian forces. Why be content with conquering Europe when there were territories to be won in North Africa, the Levant, and India? The games of territory grabbing and alliance forming were about beating the other European powers to the punch. The British Empire had India on a tether, but perhaps the British could be supplanted by the French, yielding treasures, wealth, and a strong-hold for the planned Napoleonic Empire. Equally important, the French could cut off England's prominent trade route from the Indian subcontinent to the British Isles. Egypt and Syria were the first steps for Napoleon's forces on his route to India. Is it any surprise that insects played a major role in the outcome?

Following his successful campaign in Italy, Napoleon convinced the five-person Directory (French leadership council) that they should support an invasion of Egypt and convert it into a country with a modern government that would be friendly toward France—in other words, a profitable colony. Within just a few months, Napoleon had amassed an army of 40,000, though that number also included scien-tists and engineers brought along to make the incursion more than simply a military venture, almost like a large-scale cultural adventure.

The French set sail from the coastal city of Toulon, their plans kept secret to avoid British intervention. Napoleon's invasion fleet arrived at the islands comprising Malta on June 9, 1798. After brief, if spirited, resistance at the four landing loca-tions, the French defeated the defenders. The capital of Valletta was not taken for another three days, at which point the French captured a significant number of mus-kets and artillery pieces, literally tons of gunpowder and the country's treasury. After a week, Napoleon's fleet sailed south, across the Mediterranean, but his move had been discovered by the British, although he did not realize it at the time.

The French force arrived in Egypt on July 1 and captured the fort at Alexandria the next day. Having had little rest, the troops marched south, only encountering resistance in the form of a few small skirmishes. The first engagement of the French against the Mamluks (slave and freed slave soldiers) and Arabs was on July 21 at the Battle of the Pyramids, as it became known in later years. As with many battles, this name is a slight misnomer—the Great Pyramids at Giza were not even visible from the battle site. The more numerous Mamluks and Arabs, with their scimitars and unorganized tactics, were no match for the heavily armed and disciplined French, and the French victory came later that same day. French casualties amounted to fewer than 300 wounded or killed, whereas the defenders lost 3,000–5,000 troops. The larger contingent of Mamluks retreated toward Syria. Napoleon's troops continued on to Cairo while he sent orders for his naval support ships to anchor near Alexandria, at Aboukir Bay.

While Napoleon had ventured to Malta and then to Egypt, his ships were being sought by British Admiral Horatio Nelson. On the evening of August 1, Nelson's fleet finally located the French ships anchored at Aboukir Bay. Rather than wait and attack in the morning—which would have been the conventional strategy—Nelson's ships took advantage of the poor anchoring chosen by the French ships and surprised Napoleon's Admiral de Brueys with an evening attack. Within hours, the French fleet was decimated in one of the worst naval defeats in history. The admiral's flagship, the 120-gun L'Orient, caught fire and the ship's magazine exploded, killing nearly 1,000 French sailors. In all, the French lost 11 warships, with about 1,700 killed and 1,300 wounded. By contrast, the British lost no ships, and only about 200 sailors were killed and 700 wounded. Most significant, Napoleon lost naval support for his army, which greatly altered his invasion plans.

After occupying Cairo for several months, Napoleon, fearing an attack from the east by British and Ottoman forces, led his troops to Syria. Arriving there in mid-March 1799, they laid siege to the city of Acre, the siege lasting two months. This is the same Acre against which Richard I used catapults to launch bees during the Crusades. Maybe Napoleon should have brought bees with him, because his troops could not budge the defenders of Acre.

While encamped outside the city and trying to breach its walls, an outbreak of the plague, likely caused by Oriental rat fleas, raged among Napoleon's soldiers, sharply reducing troop numbers (Chapters 4–7 discuss the plague). Napoleon allegedly believed strong moral courage could defeat the disease. Realize, the cause of the disease was unknown in 1799, but moral courage cured a lot of illnesses—at least in Napoleon's mind. Some accounts claim he visited the ill to keep morale and courage up; other accounts claim the opposite, that he was indifferent to the plight of the sick. When the siege failed, and Napoleon's troops retreated, harried by the Ottomans, the French took the plague back to Egypt with them. As the troops marched, those sickened by plague were segregated in a separate, trailing group, kept behind the main army to avoid infecting healthy troops. Being at the back of the column placed the sick troops closest to the Ottomans, who inflicted further harm on them. The fortunate ones were butchered by the Ottomans; the less fortunate were tortured by the plague until death took them. Of the 13,000 troops who arrived in Egypt with Napoleon, about 2,000 died from the plague, and half of those deaths occurred in the Syrian Campaign. Although the fleas that carried the plague were not the only reason Napoleon's invasion failed, the loss of 2,000 soldiers (this number is of deaths only—unknown numbers were sickened and too weak to fight) certainly made him realize he did not have enough remaining troops to press on. So, the campaign ended in disarray and defeat.

Score: Insects 1, Napoleon 0.

HAITI, SLAVE REVOLT, AND YELLOW FEVER

During the 1700s, Haiti was a vibrant French colony and particularly strong in scientific research. It also was the wealthiest of the colonies in the Americas, its wealth generated on the backs of African slaves who grew, harvested, and processed sugar. As we related in Chapter 9, the history of sugar in the Caribbean was written in the sweat and blood of slaves. Haiti is also relevant to our story about Napoleon.

At the onset of the 18th Century, the French West Indies Corporation had just taken control of the colony then known as Saint-Domingue (now Haiti), the western half of the island of Hispaniola. The rise of the sugar industry, plantations, and increased slave importation went hand in hand to increase the country's economic production, making it the most prosperous of France's colonies. Saint-Domingue's population swelled as more French invested in the colony. Investments in sugar saw the slave population increase greatly. At the beginning of the 18th Century, there were an estimated 5,000 slaves in Saint-Domingue; before the end of the century, the colony had 500,000 slaves. A few slaves were used for household work, and even as slave drivers, but nearly all the slaves brought from West Africa were subjected to the brutal conditions of the sugarcane fields.

Numerous factors converged to result in the slave revolt, the only successful slave revolt in history; other, more famous revolts, such as in Roman times, failed. Racist European colonists treated freed slaves and people of mixed race with disdain and as inferior. The brutal treatment of slaves, the frustration of the emancipated, and remnants of the idealism from the French Revolution sparked uprisings in the 1790s. The treatment of slaves in Saint-Domingue does seem paradoxical so soon after the French Revolution—apparently, "Liberté, Égalité, Fraternité" did not apply to slaves or people of color (however, that tripartite motto did become the motto for independent Haiti). The colony's various factions were disorganized and other European powers—the Spanish and British—tried to take advantage of the disarray and invade. But former slave, Toussaint L'Ouverture, who emerged as leader of the revolt, kept the Spanish and British from invading. Toussaint was granted power and titles of commander of Haitian forces and lieutenant governor of the colony; later, he declared himself governor general for life. With the continued uprising, Haiti—and its wealth—were slipping away from French rule.

In December 1801, Napoleon sent his brother-in-law General Charles LeClerc to put down the slave revolt. Both LeClerc and Napoleon anticipated a quick campaign, especially as 20,000 French troops were sent to add to those already in the colony. The French troops were well equipped to fight Toussaint's army and other marauders. However, they were not equipped to fight on another battle front, against an unanticipated foe—yellow fever. As discussed in Chapters 10–12, yellow fever was rampant on many Caribbean islands, especially in the swampy areas that were prime breeding grounds for *Aedes aegypti*, the mosquitos that transmit the disease. Many slaves had acquired immunity to yellow fever. However, in 1801 no one knew about the virus or immunity or that mosquitos carried the virus.

Yellow fever hit the French troops hard, especially when the rains came. In the summer of 1802, as many as 50 French troops died daily, reducing both the number and morale of the French soldiers. LeClerc negotiated a truce with L'Ouverture, but this was short-lived because of French deceit: they arrested L'Ouverture and had him shipped to France, where he died in confinement. The slaves, under new command and emboldened by the aid of their ally, yellow fever, strengthened their resolve to throw out the French. In October 1802, LeClerc became one of the disease's victims. Reinforcement troops and a replacement commander, General Rochambeau, were sent from France. The mosquitos—and the disease—meted out same unrelenting treatment to the newcomers.

The fate of the French was sealed, but still they pressed on. Finally, in November 1803, Rochambeau's depleted forces were defeated in the Battle of Vertières; he and the troops could not defeat both the slaves and the disease, and he left with the few remaining soldiers. The numbers of French troops and deaths from yellow fever vary among sources, but there were no fewer than 30,000 troop deaths, with numbers of survivors estimated between 3,000 and 7,000. Far and away, the majority died from yellow fever. Napoleon and his grand vision for the Americas were dealt a sting-ing defeat, mostly by enemies that were too small to see or recognize—yellow fever and its carrier *Aedes aegypti*. As a result, Haiti became only the second independent nation in the Americas. This defeat of the French led to another outcome, which we will return to shortly.

Score: Insects 2, Napoleon 0.

THE COLD WINTER AND LICE

Napoleon and his Grand Armée, walking west from Russia, arrived in Vilnius (in present-day Lithuania) on December 8, 1812. At that time, the Grand Armée was "Grand" in name only. Napoleon was accompanied by about 20,000 soldiers still healthy enough to fight and perhaps a similar number who were incapacitated. Twenty-thousand healthy soldiers? Less than six months before, that same Grand Armée, then *eastbound*, heading for Russia, consisted of more than 650,000 troops, then the largest army in the history of Europe. Their known enemy was the Russian army; it was their unknown enemy that caused the trouble. And in December, death followed those who could continue to march back to the west, toward France.

By 1812, the countries of Europe had been fighting in the Napoleonic Wars for nearly a decade. Arrayed against the French leader were a series of foes grouped in coalitions that were frequently dissolved and created anew. Six years before, Napoleon had mounted a naval blockade to isolate Britain and bleed the British Empire eco-nomically; Russia was then an ally of Napoleon against the British and a partner in the blockade. By 1812, Tsar Alexander and his country had been hurt economically far more than Britain, the intended target of the blockade. To mitigate the economic harm, Russia broke rank, and failed to support the French. Russia's hurt was to the tsar and the country; Napoleon's hurt was to his ego. Stung by the Russian defection, Napoleon amassed his Grand Armée for an invasion of Russia.

Assembling, feeding, clothing, and arming 650,000 troops was a feat in itself. The Grand Armée included a quarter of a million supporting troops from Germany,

Poland, Italy, and several other armies, supplementing the 400,000 soldiers from France. As part of its assembly, a network of supply depots was established from Poland into western Russia. In previous campaigns, the Grand Armée had "lived off the land," augmenting food caches with whatever they could take from the land or its occupants. Although supply depots were set up in eastern Europe, the army also relied on supply trains carrying food and water. Once in enemy territory, the supply train moved behind the army, on roads either so dusty or so filled with ruts that wagons could not keep up with the troops. Without nearby supply wagons, soldiers took whatever food they could find along the way, just to keep fed.

Almost immediately after the army crossed into Poland, the situation began to deteriorate. The peasants living in the severely impoverished region were dirty and infested with fleas and lice. The land was littered with trash and the water in wells was unsuitable for drinking. Disease began to appear, beginning with an intestinal disease, perhaps dysentery. But soon, other symptoms began to appear—rash and high fever. Napoleon's army had encountered yet another of his enemies, and it was not the Russians.

Soldiers lived in their dirty, sweaty uniforms for days and had little opportunity for washing. Troops crowded together, whether for eating or sleeping. They slept in close quarters to protect themselves from enemy attacks or from locals seeking retaliation from the soldiers' looting. Crowded conditions and poor hygiene provided ideal conditions for body lice to thrive. If any lice happened to be carrying the bacteria *Rickettsia prowazekii*, then their feces, rubbed into a wound or scratch, was all that was needed for a typhus infection to start. Cramped quarters made it all too easy for the lice to move among the soldiers, spreading the infection. Within a month of crossing into Poland, 80,000 of Napoleon's troops had died or were unable to fight—or even to march. Two weeks later, by August 25, another 25,000 had died. After each battle, the remaining Russians fell back, following General Kutuzov's masterful plan to draw French troops farther into Russia and farther from food supplies. Napoleon caught up with General Kutuzov and the Russians at Borodino on September 7 and what ensued was the bloodiest battle of the Napoleonic Wars. As with other accounts, numbers of battle casualties vary, but most estimates are about 30,000–35,000 French and 40,000–45,000 Russians dead or wounded. The French continued on their march to Moscow, but it was a greatly weakened Armée, and Moscow had been torched by the retreating Russians to eliminate food or other resources the French needed.

By October, it became clear that Alexander would not surrender. Autumn was well advanced, and winter was on its way. Napoleon and his depleted troops began to retreat on October 19, hoping to find shelter before deep winter set in. The road of retreat was soon littered with the bodies of dead French soldiers. Weak and hungry, retreating soldiers died while marching or collapsed, too weak to carry on. Survivors took what clothing they could from the dead to try to keep warm. The clothing taken from any soldier who had died from typhus likely contained infected lice, which would then sicken the next soldier to wear it. In this manner, lice maintained their hegemony over the no-longer-Grand Armée. The depleted Armée arrived in Vilnius on December 8, but lice and typhus were still present among the weakened, hungry troops. Thousands more would die before the troops and stragglers could re-cross

Figure 14.1 *Retreat from Russia*, by Pryanishnikov.
Source: Dennis Jarvis via Flickr/CC BY-SA 2.0

the Niemen River, out of Russian territory; the sick were left behind to die when the healthy marched on (Figure 14.1).

During excavations in Vilnius in 2001, mass graves were found: nearly 3,000 skeletons were identified as French soldiers by fragments of uniforms or their buttons. Analyses of teeth and other remains showed that nearly 30% of the individuals assayed had remnants of *R. prowazekii* and *Bartonella quintana*, the causative agent of trench fever; both are transmitted by body lice.

Historically, blame for Napoleon's defeat has been pinned on the winter cold, the lack of a complete supply train to feed soldiers, or the Russian Army. Napoleon's troops could have overcome most of the privations, but typhus carried by body lice is unforgiving. The victors? General Kutuzov, yes. General Typhus and General Lice, most certainly.

Score: Insects 3, Napoleon 0

NEVER DOUBT FOR A MOMENT

Three of the insects we presented in this book defeated Napoleon, not together, but on independent occasions and in different places. If the insects had not been there, history may well have moved in different directions. The stories all are connected, not just through Napoleon, but along the Silk Roads or its extensions. The silken thread that runs through the insects and history is there, once we look a little deeper into the history that we didn't learn or that was not included in history books. The stories are real, and insects played a major part in them.

There is another twist to the story of Napoleon—actually, more than one. The symbol for Napoleon's empire was the honey bee, another of our five insects. Napoleon adopted the honey bee as his symbol because its ancient history pre-dated the fleur-de-lys, the symbol of the French monarchy deposed in the revolution. One Napoleonic supporter suggested the emblem should be a hive of swarming bees and it should be emblazoned on every government building. Another opinion offered was that bees could not be used because the bees serve a queen. Napoleon ignored the advice and used the bees as his symbol. It was not the first time he ignored advice. Bees were embroidered on all of his coronation gowns (made of silk) and were depicted on clothing, glass, and furniture. Bees were used as symbols for the flag he created for Elba, his home in exile, and the velvet mantle used in his funeral cortege was covered by embroidered bees. Counting bees, that actually makes four of our insects connected to Napoleon. To make the story complete, Napoleon's uniform did include silk stockings and a silk collar, and the outfit he wore for his 1804 coronation as Emperor of France included white silk breeches, a white silk tunic, and satin robes adorned with embroidered bees. Five out of five insects significantly interacting with him is not bad, especially for such a prominent figure in history.

One final twist in the Napoleonic tales resonates especially with us, and we will let you in on it, too. We take you back to Haiti and the departure of the French troops in the wake of their battle losses to yellow fever and *Aedes aegypti*. In 1801, Napoleon had plans for a great empire in North America. After the decimation of his troops in Saint-Dominique, Napoleon aborted his grand plan, and the French abandoned their colonies in the Americas. In 1803, US President Thomas Jefferson was interested in buying the city of New Orleans from the French. The French made a counter offer of New Orleans plus the vast lands that stretched westward from the Mississippi River. What a deal! Buy New Orleans and we'll throw in 2.14 million square kilometers of productive land! The $15 million transaction, called the Louisiana Purchase, doubled the size of a very young United States, led to the country's robust growth and becoming a global power, and the land transfer kept the French from reaching similar global strength. All of that came about because of a mosquito. One reason we are not writing this book in French is that mosquito.

So, we end with our repeated admonition: Never doubt for a moment that insects have affected human history. Because they have, and they all have tales to tell. Zillions of tiny, six-legged hammers looking for nails.

BIBLIOGRAPHY

SECTION 1. SILK AND SILKWORM

"Agricultural Revolution—China 6,000BC." https://sites.ualberta.ca/~vmitchel/rev2. html. Accessed March 21, 2020.

Balter, M. 2005. "The Seeds of Civilization." https://www.smithsonianmag.com/history/ the-seeds-of-civilization-78015429/. Accessed September 18, 2020.

Beckwith, C. I. 2011. *Empires of the Silk Road: A History of Central Eurasia from the Bronze Age to the Present*. Princeton, NJ: Princeton University Press.

Black, E. 2001. *IBM and the Holocaust: The Strategic Alliance between Nazi Germany and America's Most Powerful Corporation*. New York: Crown Publishers.

Brite, E. B., and J. M. Marston. 2016. "Environmental Change, Agricultural Innovation, and the Spread of Cotton Agriculture in the Old World." *Journal of Anthropological Archaeology* 32: 39–53.

Britannica Online Encyclopedia. "Ramie." https://www.britannica.com/print/article/ 490642. Accessed August 22, 2020.

Byzantine Military Blogspot. 2013. "The Byzantine Silk Industry." https://byzantinemili-tary.blogspot.com/2013/02/the-byzantine-silk-industry.html. Accessed September 3, 2020.

"Canvas." https://www.lexico.com/en/definition/canvas. Accessed March 21, 2020.

Cartwright, M. 2017. "Silk in Antiquity." *World History Encyclopedia*.https://www.ancient. eu/Silk/. Accessed March 21, 2020.

Connecting Women to the Silk Road. "Wearing Silk, Popularity and Prohibitions." http:// womeninworldhistory.com/silk-road-03.html. Accessed November 9, 2020.

"Cotton." https://www.lexico.com/en/definition/cotton. Accessed March 21, 2020.

Frankopan, P. 2015. *The Silk Roads: A New History of the World*. New York: Alfred A. Knopf.

"Han Dynasty." https://www.history.com/topics/ancient-china/han-dynasty. Accessed March 14, 2020.

Harper, D. "Line." Online Etymology Dictionary. Accessed March 21, 2020.

"Hemp." https://www.lexico.com/en/definition/hemp. Accessed March 21, 2020.

Herrera, R. J., and R. Garcia-Bertrand. 2018. *Ancestral DNA, Human Origins, and Migrations*. New York: Academic Press.

History.com Editors. 2019. https://www.history.com/topics/pre-history/neolithic-revolution. Accessed March 21, 2020.

History of Clothing.com. "History of Wool Making." http://www.historyofclothing.com/ textile-history/wool-history/. Accessed March 9, 2020.

"History of Silk." http://silkroadfoundation.org/artl/silkhistory.shtml. Accessed March 9, 2020.

"History of Silk Fabric." http://www.historyofclothing.com/textile-history/history-of-silk/. Accessed March 9, 2020.

"History of the Silk Road." http://www.silk-road.com/artl/silkhistory.shtml. Accessed February 21, 2020.

Landry, D. 2013. "History of Silk Production." http://www.mansfieldct-history.org/history-of-silk-production/. Accessed March 9, 2020.

Lerner, J. A., and T. Wide. "Who Were the Sogdians?" https://sogdians.si.edu/introduction/. Accessed August 27, 2020.

"Linen." https://www.lexico.com/en/definition/linen. Accessed March 9, 2020.

Maestri, N. "The Domestication History of Cotton (*Gossypium*)." ThoughtCo. https://www.thoughtco.com/domestication-history-of-cotton-gossypium-170429. Accessed March 21, 2020.

Moulherat, C., M. Tengberg, J.-F. Haquet, and B. Mille. 2002. "First Evidence of Cotton at Neolithic Mehrgarh, Pakistan: Analysis of Mineralized Fibres from a Cotton Bead." *Journal of Archaeological Science* 29: 1393–1401.

Orendi, A. 2019. "Flax Cultivation in the Southern Levant and Its Development during the Bronze and Iron Age." *Quaternary International*. https://doi.org/10.1016/j.quaint.2019.10.007.

Organiccotton.org. "The History of Cotton Production." https://organiccotton.org/oc/Cotton-general/World-market/History-of-cotton.php. Accessed March 9, 2020.

"Programming Patterns: The Story of the Jacquard Loom." 2019. https://www.scienceandindustrymuseum.org.uk/objects-and-stories/jacquard-loom. Accessed September 6, 2020.

Renny-Benfield, S., J. T. Page, J. A. Udall, W. S. Sanders, D. G. Peterson, M. A. Arick II, C. E. Grover, and J. F. Wendel. 2016. "Independent Domestication of Two Old World Cotton Species." *Genome Biology and Evolution* 8: 1940–47.

"Sick or Silk: How Silkworms Spun The Germ Theory Of Disease." 2019. American Society for Microbiology. https://asm.org/Articles/2019/December/Sick-or-Silk-How-Silkworms-Spun-the-Germ-Theory-of. Accessed October 8, 2020.

"Silk Production and Trade in Medieval Times." https://www.thoughtco.com/silk-lustrous-fabric-1788616. Accessed September 3, 2020.

"Silk Revolution of the Middle Ages." https://www.ancient.eu/Silk/. Accessed March 9, 2020.

Steinhaus, E. 1975. "Disease in a Minor Chord: Being a Semihistorical and Semibiographical Account of a Period in Science When One Could Be Happily Yet Seriously Concerned with the Diseases of Lowly Animals without Backbones, Especially the Insects." https://kb.osu.edu/handle/1811/29317. Accessed October 8, 2020.

Textile School. "History of Textile Fabrics." https://www.textileschool.com/130/history-of-textile-fabrics/. Accessed March 8, 2020.

United States Census Bureau. "Herman Hollerith." https://www.census.gov/history/www/census_then_now/notable_alumni/herman_hollerith.html. Accessed September 3, 2020.

Yingshu, Yu. 1967. *Trade and Expansion in Han China. A Study in the Structure of Sino-Barbarian Economic Relations*. Berkeley: University of California Press.

SECTION 2. ORIENTAL RAT FLEA AND THE PLAGUE

"25 Million Dead." https://nationalinterest.org/blog/buzz/25-million-dead-how-black-death-nearly-destroyed-europe-91161. Accessed June 20, 2020.

Alibek, K., and K. Handelman. 1999. *Biohazard: The Chilling True Story of the Largest Covert Biological Weapons Program in the World—Told from the Inside by the Man Who Ran It*. New York: Random House.

Barbieri, R., M. Drancourt, and D. Raoult. 2019. "Plague, Camels, and Lice." *Proceedings of the National Academy of Sciences USA* 116 (16): 7620–7621.

Ben Ari, T., S. Neerinckx, K. L. Gage, K. Kreppel, A. Laudisoit, H. Leirs, and N. Chr. Stenseth. 2011. "Plague and Climate: Scales Matter." *PLoS Pathog* 7 (9): e1002160. https://doi.org/10.1371/journal.ppat.1002160.

Benavides-Montaño, J. A., and V. Vadyvaloo. 2017. "*Yersinia pestis* Resists Predation by *Acanthamoeba castellanii* and Exhibits Prolonged Intracellular Survival." *Applied and Environmental Microbiology* 83: e00593–17. doi:10.1128/AEM.00593-17.

Bertherat, E. 2016. Plague around the world, 2010–2015. Weekly Epidemiological Record (World Health Organization) 91:89–93.

Bramanti, B., A. Namouchi, B. V. Schmid, K. R. Dean, and N. Chr. Stenseth. 2019. "Out of the Land of Darkness: Plague on the Fur Trade Routes." *Proceedings of the National Academy of Sciences USA* 116 (16): 7620–7621.

Bramanti, B., K. R. Dean, L. Walloe, and N. Chr. Stenseth. 2019. "The Third Plague Pandemic in Europe." *Proceedings of the Royal Society B* 286: 20182429.

Buckland, P. C., and J. P. Sadler. 1989. "A Biogeography of the Human Flea, *Pulex irritans* L. (Siphonaptera: Pulicidae)." *Journal of Biogeography* 16: 115–120.

Dean, K. R., F. Krauer, L. Walloe, O. C. Lingjaerde, B. Bramanti, N. Chr. Stenseth, and B. V. Schmid. 2018. "Human Ectoparasites and the Spread of Plague in Europe during the Second Pandemic." *Proceedings of the National Academy of Sciences USA* 115 (6): 1304–1309.

Demeure, C. E., O. Dussurget, G. M. Fiol, A.-S. Le Guern, C. Savin, and J. Pizarro-Cerda. 2019. "*Yersinia pestis* and Plague: An Updated View on Evolution, Virulence Determinants, Immune Subversion, Vaccination, and Diagnostics." *Genes & Immunity* 20: 357–370.

Ditrich, H. 2017. "The Transmission of the Black Death to Western Europe—A Critical Review of the Existing Evidence." *Mediterranean Historical Review* 32 (1): 25–39.

Eisen, R. J., J. N. Borchert, J. L. Holmes, G. Amatre, K. Van Wyk, R. E. Enscore, N. Babi, L. A. Atiku, A. P. Wilder, S. M. Vetter, S. W. Bearden, J. A., Montenieri, and K. L. Gage. 2008. "Early-Phase Transmission of *Yersinia pestis* by Cat Fleas (*Ctenocephalides felis*) and Their Potential Role as Vectors in a Plague-Endemic Region of Uganda." *American Journal of Tropical Medicine and Hygiene* 78 (6): 949–956.

Fenyuk, B. K. 1960. "Experience in the Eradication of Enzootic Plague in the North-West Part of the Caspian region of the USSR." *Bulletin of the World Health Organization* 23 (2–3): 263–273. PMID: 13822193; PMCID: PMC2555585.

Gräslund, B., and N. Price. 2012. "Twilight of the Gods? The 'Dust Veil Event' of AD 536 in Critical Perspective." *Antiquity* 86: 428–443. doi:10.1017/S0003598X00062852.

Harper, K. 2017. *The Fate of Rome: Climate, Disease, and the End of an Empire*. Princeton, NJ: Princeton University Press.

Hester, J. L. 2020. "Sold: Isaac Newton's Notes about the Bubonic Plague." https://www.atlasobscura.com/articles/isaac-newton-plague. Accessed June 20, 2020.

Hinnebusch, B. J. 2012. "The Evolution of Flea-Borne Transmission in *Yersinia pestis*." *Current Issues in Molecular Biology* 7: 197–212.

Hinnebusch, B. J., C. O. Jarrett, and D. M. Bland. 2017. "'Fleaing' the Plague: Adaptations of *Yersinia pestis* to Its Insect Vector that Lead to Transmission." *Annual Review of Microbiology* 71: 215–232.

History.com Editors. 2020. "Black Death—Causes, Symptoms & Impact." www.history.com/topics/middle-ages/black-death. Accessed May 16, 2020.

Hufthammer, A. K., and L. Walloe. 2013. Rats cannot have been intermediate hosts for *Yersinia pestis* during medieval plague epidemics in Northern Europe. *Journal of Archaeological Science* 40:1752–1759.

Jones, S. D., B. Atshabar, B. V. Schmid, M. Zuk, A. Amramina, and N. Chr. Stenseth. 2019. "Living with Plague: Lessons from the Soviet Union's Antiplague System." *Proceedings of the National Academy of Sciences USA* 116: 9155–9163.

Keller, M., M. A. Spyrou, C. L. Scheib, G. U. Neumann, A. Kröpelin, B. Haas-Gebhard, B. Päffgen, J. Haberstroh, A. Ribiera i Lacomba, C. Raynaud, C. Cessford, R. Durand, P. Stadler, K. Nägele, J. S. Bates, B. Trautmann, S. A. Inskip, J. Peters, J. E. Robb, T. Kivisild, D. Castex, M. McCormick, K. I. Bos, M. Harbeck, A. Herbig, and J. Krause. 2019. "Ancient *Yersinia pestis* Genomes from across Western Europe Reveal Early Diversification during the First Pandemic (541–750)." *Proceedings of the National Academy of Sciences USA* 116: 12363–12372.

Lawrence, A. L., C. E. Webb, N. J. Clark, A. Halajian, A. D. Mihalca, J. Miret, G. D'Amico, G. Brown, B. Kumsa, D. Modry, and J. Slapeta. 2019. "Out-of-Africa, Human-Mediated Dispersal of the Common Cat Flea, *Ctenocephalides felis*: The Hitchhiker's Guide To World Domination." *International Journal for Parasitology* 49: 321–336.

Mahoney, C. W., J. W. Toppin, and R. A. Zilinskas (eds.). 2013. "Stories of the Soviet Anti-Plague System." James Martin Center for Nonproliferation Studies, Middlebury Institute for International Studies, Monterey, California. CNS Occasional Paper #18, September 2013. http://www.nonproliferation.org/wp-content/uploads/2017/10/op18-soviet-antiplague.pdf.

Maugh, T. "An Empire's Epidemic; Scientists Use DNA in Search for Answers to 6th Century Plague." https://www.ph.ucla.edu/epi/bioter/anempiresepidemic.html. Accessed May 16, 2020.

Mizokami, K. "The Black Death: How Rats, Fleas and Germs Almost Wiped Out Europe." https://nationalinterest.org/blog/the-black-death-how-rats-fleas-germs-almost-wiped-out-europe-19745. Accessed June 20, 2020.

Mordechai, L., and M. Eisenberg. 2019. "Rejecting Catastrophe: The Case of the Justinianic Plague." *Past & Present* 244 (1): 3–50.

Moskowitz, C. 2008. "Black Death Was Selective." https://www.livescience.com/9561-black-death-selective.html. Accessed June 22, 2020.

Namouchi, Q., M. Guellil, O. Kersten, S. Hänsch, C. Ottoni, B. V. Schmid, E. Pacciani, L. Quaglia, M. Vermunt, E. L. Bauer, M. Derrick, A. O. Jensen, S. Kacki, S. K. Cohn Jr., N. C. Stenseth, and B. Bramanti. 2018. "Integrative Approach Using *Yersinia pestis* Genomes to Revisit the Historical Landscape of Plague during the Medieval Period." *Proceedings of the National Academy of Sciences USA* 115: E11790–E11797.

Parry, W. 2011. "Molecular Clues Hint at What Really Caused the Black Death." https://www.livescience.com/15937-black-death-plague-debate.html. Accessed June 22, 2020.

Perry, R. D., and J. D. Fetherston. 1997. "*Yersinia pestis*—Etiologic Agent of Plague." *Clinical Microbiology Reviews* 10: 35–66. doi:10.1128/CMR.10.1.35-66.1997. PMID:8993858; PMCID:PMC172914.

Puckett, E. E., D. Orton, and J. Munshi-South. 2020. "Commensal Rats and Humans: Integrating Rodent Phylogeography and Zooarchaeology to Highlight Connections between Human Societies." *BioEssays* 42: 1900160. doi:10.1002/bies.201900160.

Rasmussen, S., M. E. Allentoft, K. Nielsen, L. Orlando, M.Sikora, K.-G. Sjögren, A. G. Pedersen, M. Schubert, A. Van Dam, C. M. O. Kapel, H. B. Nielsen, S. Brunak, P. Avetisyan, A. Epimakhov, M. V. Khalyapin, A. Gnuni, A. Kriiska, I Lasak, M. Metspalu, V. Moiseyev, A. Gromov, D. Pokutta, L. Saag, L. Varul, L. Yepiskoposyan, T. Sicheritz-Pontén, R. A. Foley, M. M. Lahr, R. Nielsen, K. Kristiansen, and E.

Willerslev. 2015. "Early Divergent Strains of *Yersinia pestis* in Eurasia 5,000 Years Ago." *Cell* 163: 571–582.

Rosen, W. 2007. *Justinian's Fleas*. New York: Penguin Books.

Snell, M. 2019. "The Sixth-Century Plague." https://www.thoughtco.com/the-sixth-century-plague-1789291. Accessed May 16, 2020.

Spyrou, M. A., M. Keller, R. I. Tukhbatova, C. L. Scheib, E. A. Nelson, A. A. Valtuena, G. U. Neumann, D. Walker, A. Alterauge, N. Carty, C. Cessford, H. Fetz, M. Gourvennec, R. Hartle, M. Henderson, K. von Heyking, S. A. Inskip, S. Kacki, F. M. Key, E. L. Knox, C. Later, P. Maheshwari-Aplin, J. Peters, J. E. Robb, J. Schreiber, T. Kivisild, D. Castex, S. Losch, M. Harbeck, A. Herbig, K. I. Bos, and J. Krause. 2019. "Phylogeography of the Second Plague Pandemic Revealed through Analysis of Historical *Yersinia pestis* genomes." *Nature Communications* 10: 4470. https://doi.org/10.1038/s41467-019-12154-0.

Spyrou, M. A., R. I. Tukhbatova, C.-C. Wang, A. A. Valtuena, A. K. Lankapalli, V. V. Kondrashin, V. A. Tsybin, A. Khokhlov, D. Kuhnert, A. Herbig, K. I. Bos, and J. Krause. 2018. "Analysis of 3800-Year-Old *Yersinia pestis* Genomes Suggests Bronze Age Origin for Bubonic Plague." *Nature Communications* 9: 2234. doi:10.1038/s41467-018-04550-9.

Thompson, H. "Bubonic Plague Family Tree Sheds Light on the Risks of New Outbreaks." https://www.smithsonianmag.com/science-nature/bubonic-plague-family-tree-sheds-light-on-risk-of-new-outbreaks-180949498/. Accessed May 16, 2020.

"Weapons of Mass Destruction." https://www.globalsecurity.org/wmd/intro/bio_plague.htm. Accessed June 20, 2020.

White, L. A., and L. Mordechai. 2020. "Modeling the Justinianic Plague: Comparing Hypothesized Transmission Routes." *PLoS ONE* 15 (4): e0231256.

Xu, L., L. C. Stige, H. Leirs, S. Neerinckx, K. L. Gage, R. Yang, Q. Liu, B. Bramanti, K. R. Dean, H. Tang, Z. Sun, N. Chr. Stenseth, and Z. Zhang. 2019. "Historical and Genomic Data Reveal the Influencing Factors on Global Transmission Velocity of Plague during the Third Pandemic." *Proceedings of the National Academy of Sciences USA* 116: 11833–11838.

Zhang, S. 2019. "Soviets Tried So, So Hard to Eliminate the Plague." https://www.theatlantic.com/science/archive/2019/05/when-soviets-tried-to-eradicate-the-plague/589570/. Accessed August 17, 2020.

Zhao, F., T. Zhang, J. Su, Z. Huang, A. Wu, and G. Lin. 2018. "Genetic Differentiation of the Oriental Rat Flea, *Xenopsylla cheopis*, from Two Sympatric Host Species." *Parasites & Vectors* 11: 343. https://doi.org/10.1186/s13071-018-2903-8.

Zurita, A., R. Callejon, A. M. Garcia-Sanchez, M. Urdapilleta, M. Lareschi, and C. Cutillas. 2019. "Origin, Evolution, Phylogeny and Taxonomy of *Pulex irritans*." *Medical and Veterinary Entomology* 33: 296–311.

SECTION 3. LICE IN WAR AND PEACE

"A Brief History of the English Prison System." 2007. https://h2g2.com/approved_entry/A20961137#:~:text=%20A%20Brief%20History%20Of%20The%20English%20Prison,dramatically%20when%20the%20Criminal%20Justice%20Act...%20More%20. Accessed August 12, 2020.

Allen, A. 2014. "Killing in the Name of Cleanliness: The Shocking Stupidity and Brutality of Nazi Doctors' Response to Typhus." Slate.com. https://nationalpost.com/news/killing-in-the-name-of-cleanliness-the-shocking-stupidity-and-brutality-of-nazi-doctors-response-to-typhus. Accessed August 12, 2020.

Amanzougaghene, N., F. Fenollar, D. Raoult, and O. Mediannikov. 2020. "Where Are We with Human Lice? A Review of the Current State of Knowledge." *Frontiers in Cellular and Infectious Microbiology* 9: 474. doi:10.3389/fcimb.2019.00474.

Andersson, J. O., and S. G. E. Andersson. 2000. "A Century of Typhus, Lice and *Rickettsia*." *Research in Microbiology* 151: 143–150.

Anstead, G. M. 2016. "The Centenary of the Discovery of Trench Fever, An Emerging Infectious Disease of World War 1." *Lancet Infectious Diseases* 16: e164–72. http://dx.doi.org/10.1016/S1473-3099(16)30003-2.

Bitto, R. 2020. "Carmelita Torres and the 1917 Bath Riots." https://mexicounexplained.com/carmelita-torres-and-the-1917-bath-riots. Accessed September 20, 2020.

Brouqui, P. 2011. "Arthropod-Borne Diseases Associated with Political and Social Disorder." *Annual Review of Entomology* 56: 357–374.

Burgess, I. F. 2004. "Human Lice and Their Control." *Annual Review of Entomology* 49: 457–481.

Editors of the Encyclopedia Britannica. 2020. "Howard T. Ricketts." https://www.britannica.com/biography/Howard-T-Ricketts. Accessed September 20, 2020.

Editors of the Encyclopedia Britannica. 2009. "Sucking Louse." https://www.britannica.com/animal/sucking-louse. Accessed September 20, 2020.

Gallagher, J. A. 1936. "The Irish Emigration of 1847 and Its Canadian Consequences." *CCHA (Canadian Catholic Historical Association) Report* 3: 43–57.

Gelston, A. L., and T. C. Jones. 1977. "Typhus Fever: Report of an Epidemic in New York City in 1847." *The Journal of Infectious Diseases* 136: 813–821.

George, R. 1920. "When Typhus Raged in Canada." *The Public Health Journal* 11: 548–551. http://www.jstor.org/stable/41972635.

Holliday, T. K. 2020. "What an 1836 Typhus Outbreak Taught the Medical World about Epidemics. https://www.smithsonianmag.com/history/what-1836-typhus-outbreak-taught-medical-world-about-epidemics-180974707/#:~:text=What%20an%201836%20Typhus%20Outbreak%20Taught%20the%20Medical,and%20typhoid%20was%20American%20doctor%20William%20Wood%20Gerhard. Accessed September 2, 2020.

Howard League for Penal Reform. "History of the Penal System." https://howardleague.org/history-of-the-penal-system/. Accessed August 12, 2020.

Keating, J. 1996. *Irish Famine Facts*. Oak Park, Carlow, Ireland: Teagasc (Agricultural and Food Development Authority).

Kittler, R., M. Kayser, and M. Stoneking. 2003. "Molecular Evolution of *Pediculus humanus* and the Origin of Clothing." *Current Biology* 13: 1414–1417.

Kittler, R., M. Kayser, and M. Stoneking. 2004. "Erratum. Molecular evolution of *Pediculus humanus* and the Origin of Clothing." *Current Biology* 14: 2309.

Laxton, E. 1998. *The Famine Ships: The Irish Exodus to America*. New York: Henry Holt and Company.

Patterson, K. D. 1993. "Typhus and Its Control in Russia, 1870–1940." *Medical History* 37: 361–381.

Raoult, D., and V. Roux. 1999. "The Body Louse as a Vector of Reemerging Human Diseases." *Clinical Infectious Diseases* 29: 888–911.

Sexton, C. 2020. "How a Public Health Campaign in the Warsaw Ghetto Stemmed the Spread of Typhus." https://www.smithsonianmag.com/science-nature/how-a-public-health-campaign-warsaw-ghetto-stemmed-spread-typhus-180975418/. Accessed August 12, 2020.

Tschanz, D. W. "Typhus Fever on the Eastern Front in World War II." https://www.montana.edu/historybug/wwi-tef.html#:~:text=Typhus%20Fever%20on%20the%20

Eastern%20Front%20in%20World,resulted%20from%20miasma%20or%20 an%20imbalance%20in%20humors. Accessed September 1, 2020.

"Typhus Fevers." 2019. U. S. Centers for Disease Control and Prevention. https://www.cdc. gov/typhus/index.html. Accessed September 19, 2020.

"Typhus in World War I." 2014. Microbiology Society. https://microbiologysociety.org/ publication/past-issues/world-war-i/article/typhus-in-world-war-i.html. Accessed September 1, 2020.

Woodham-Smith, Cecil. 1991. *The Great Hunger: Ireland 1845–1849*. New York: Penguin Books.

SECTION 4. *AEDES AEGYPTI* AND YELLOW FEVER

"Arkansas: Yellow Fever." 2004. Transcribed by T. Easley. http://www.usgennet.org/usa/ ar/county/greene/arkansasyellowfever.htm. Accessed January 11, 2020.

Barnet, E. B. 1915. "Dr. Carlos J. Finlay, Discoverer of the Theory of the Transmission of Yellow Fever by the Mosquito." *California State Journal of Medicine* 13 (12): 476–478.

Bennett, I. E. 1915. "History of the Panama Canal." Washington, DC: Historical Publishing Company. http://www.czbrats.com/Menus/bennett.htm.

Brink, S. 2016. "Yellow Fever Timeline: The History of a Long Misunderstood Disease." https://www.npr.org/sections/goatsandsoda/2016/08/28/491471697/yellow- fever-timeline-the-history-of-a-long-misunderstood-disease. Accessed January 20, 2020.

British Broadcasting Corporation. "British Colonies." https://www.bbc.co.uk/bitesize/ guides/zjyqtfr/revision. Accessed January 22, 2020.

Bryant, J. E., E. C. Holmes, and A. D. T. Barrett. 2007. "Out of Africa: A Molecular Perspective on the Introduction of Yellow Fever Virus into the Americas." *PLoS Pathogens* 3 (5): e75. doi:10.1371/journal.ppat.0030075.

Canal, D. 2016. "Panama Canal—Facts & Summary." http://www.history.com/topics/ panama-canal. Accessed January 20, 2020.

"Carlos J. Finlay." 2019. Encyclopedia Britannica. https://www.britannica.com/biography/ Carlos-J-Finlay. Accessed January 20, 2020.

Cavanaugh, D. "Finlay's Theory." http://news.hsl.virginia.edu/index.html%3Fp=7122. html. Accessed January 20, 2020.

Cavanaugh, D. "The Havana Commission." http://news.hsl.virginia.edu/index. html%3Fp=7105.html. Accessed January 20, 2020.

Cavanaugh, D. 2019. "Walter Reed and the Scourge of Yellow Fever." UVA Today: Research & Discovery. https://news.virginia.edu/content/walter-reed-and-scourge-yellow- fever. Accessed February 2, 2020.

Cavanaugh, D. "Yellow Fever in the United States." http://news.hsl.virginia.edu/index. html%3Fp=7053.html. Accessed January 20, 2020.

Chippaux, J.-P., and A. Chippaux. 2018. "Yellow Fever in Africa and the Americas: A Historical and Epidemiological Perspective." *Journal of Venomous Animals and Toxins Including Tropical Diseases* 24: 20. https://doi.org/10.1186/s40409-018-0162-y.

Costa, A., V. C. Smith, G. Macedonio, and N. E. Matthews. 2014. "The Magnitude and Impact of the Youngest Toba Tuff Super-Eruption." *Frontiers in Earth Science*. https://www.frontiersin.org/article/10.3389/feart.2014.00016. DOI=10.3389/ feart.2014.00016.

Crosby, M. C. 2006. *The American Plague: The Untold Story of Yellow Fever, the Epidemic that Shaped Our History*. New York: Berkley Books.

"Cuba and the Slave Trade." http://www.tracesofthetrade.org/guides-and-materials/his- torical/cuba-and-the-slave-trade/. Accessed January 22, 2020.

Cunningham, R. D. "The Black 'Immune' Regiments in the Spanish-American War." https://armyhistory.org/the-black-immune-regiments-in-the-spanish-american-war/. Accessed January 20, 2020.

Donnella, L. 2018. "How Yellow Fever Turned New Orleans into the 'City of the Dead.'" https://www.npr.org/sections/codeswitch/2018/10/31/415535913/how-yellow-fever-turned-new-orleans-into-the-city-of-the-dead. Accessed January 11, 2020.

Duffy, J. 1968. "Yellow Fever in the Continental United States during the 19th Century." *Bulletin of the New York Academy of Medicine* 44: 687–701.Evans, D. L. 2020. "A Geophysical and Climatological Assessment of New Guinea—Implications for the Origins of *Saccharum*." *BioRxiv*. https://doi.org/10.1101/2020.06.20.162842.

Evans, D. L., and S. V. Joshi. 2020. "On the Validity of the *Saccharum* Complex and the Saccharinae Subtribe: A Re-Assessment." *BioRxiv*. https://doi.org/10.1101/2020.07.29.22675.

Evans, D. L., and S. V. Joshi. 2020. "Origins, History and Molecular Characterization of Creole Cane." *BioRxiv*. https://doi.org/10.1101/2020.07.29.226852.

Feng, P. 2015. "Major Walter Reed and the Eradication of Yellow Fever." https://armyhistory.org/major-walter-reed-and-the-eradication-of-yellow-fever/. Accessed January 20, 2020.

Greenfield, S. M. 1977. "Madeira and the Beginnings of New World Sugar Cane Cultivation And Plantation Slavery, in V. Rubin and A. Tuden, eds. *Comparative Perspectives on Slavery in New World Plantation Societies.*" *Annals of the New York Academy of Science* 292: 536–552.

Greenfield, S. M. 1979. "Sugar Cane and Slavery. Historical Reflections." *Current Directions in Slave Studies* 6 (1): 85–119.

Harrington, L. C., John D. Edman, and T. W. Scott. 2001. "Why Do Female *Aedes aegypti* (Diptera: Culicidae) Feed Preferentially and Frequently on Human Blood?" *Journal of Medical Entomology* 38: 411–422.

"Historical Statistics of the United States, Colonial Times–1957." https://www2.census.gov/library/publications/1960/compendia/hist_stats_colonial-1957/hist_stats_colonial-1957-chQ.pdf. Accessed January 11, 2020.

History.com Editors. 2019. "Spanish-American War." https://www.history.com/topics/early-20th-century-us/spanish-ameri. Accessed January 15, 2020.

History.com Editors. 2018. "Suez Canal." https://www.history.com/topics/africa/suez-canal. Accessed January 20, 2020.

Inikori, J. E. 2002. *Africans and the Industrial Revolution in England: A Study in International Trade and Economic Development.* Cambridge University Press, Cambridge.

Jennings R. G. 1878. "The Quarantine at Little Rock, Arkansas, during August, September, and October, 1878, against the Yellow Fever Epidemic in Memphis and the Mississippi Valley." *Public Health Papers and Reports* 4: 223–227.

Kaur, H. 2020. A New DNA Study Offers Insight into the Horrific Story of the Trans-Atlantic Slave Trade. https://www.cnn.com/2020/07/26/us/dna-transatlantic-slave-trade-study-scn-trnd/index.html. Accessed July 26, 2020.

Keith, J. 2012. *Fever Season: The Story of a Terrifying Epidemic and the People Who Saved a City.* New York: Bloomsbury Press.

Lockley, T. J. 2013. "Black Mortality in Antebellum Savannah." *Social History of Medicine* 26: 633–652.

Martinez, J. M. 2012. *Terrorist Attacks on American Soil: From the Civil War Era to the Present.* Lanham, MD: Rowman and Littlefield.

McCarthy, M. 2001. "A Century of the US Army Yellow Fever Research." *Lancet* 357: 1772.

Moore, J. W. 2009. "Madeira, Sugar, and the Conquest of Nature in the 'First' Sixteenth Century: Part I: From Island of Timber to Sugar Revolution, 1420–1506." *Review (Fernand Braudel Center)* 32 (4): 345–390.

Moore, J. W. 2010. "Madeira, Sugar, and the Conquest of Nature in the 'First' Sixteenth Century: Part II: From Regional Crisis to Commodity Frontier, 1506–1530." *Review (Fernand Braudel Center)* 33 (1): 1–24.

Moya Pons, F. 2007. *History of the Caribbean: Plantations, Trade, and War in the Atlantic World*. Princeton, NJ: Markus Wiener Publishers.

National Archives (UK). "Africa and the Caribbean: The Caribbean and Trade." https://www.nationalarchives.gov.uk/pathways/blackhistory/africa_caribbean/caribbean_trade.htm. Accessed January 22, 2020.

"New Tastes, New Trades." On the Water. Smithsonian Institution. https://americanhistory.si.edu/onthewater/exhibition/1_3.html. Accessed January 22, 2020.

"Panama Scandal." 2016. Encyclopedia Britannica. https://www.britannica.com/event/Panama-Scandal. Accessed January 20, 2020.

Patterson, R. 1989. "Dr. William Gorgas and His War with the Mosquito." *Canadian Medical Association Journal* 141: 596–598.

"Port of Havana." http://www.worldportsource.com/ports/review/CUB_Port_of_Havana_1859.php. Accessed January 22, 2020.

Public Broadcasting Service. 2020. "How Walter Reed Earned His Status as a Legend and Hospital Namesake." *PBS News Hour: Health*. https://www.pbs.org/newshour/health/walter-reed-earned-status-legend-hospital-namesake. Accessed January 20, 2020.

"Quarantine." www.etymonline.com/word/quarantine. Accessed March 4, 2020.

Ray, M. 2019. "William Crawford Gorgas." Encyclopedia Britannica. https://www.britannica.com/biography/William-Crawford-Gorgas. Accessed January 20, 2020.

Rucker, W. C. 1906. "The Epidemiology of Yellow Fever." *Journal of the Royal Army Medical Corps* 7: 494–497. doi:10.1136/jramc-07-05-07.

Schulz, M. G. 2009. "Henry Rose Carter." *Emerging Infectious Diseases* 15: 1682–1684.

"Sugar Boom." http://cubahistory.org/en/sugar-boom-a-slavery/sugar-boom.html. Accessed January 22, 2020.

"T. R. Roosevelt and the Panama Canal." http://www.kumc.edu/school-of-medicine/history-and-philosophy-of-medicine/clendening-history-of-medicine-library/special-collections/panama-canal/tr-roosevelt-and-the-panama-canal.html. Accessed January 22, 2020.

"The French Canal Construction." http://www.pancanal.com/eng/history/history/french.html. Accessed January 22, 2020.

"The Great Fever." Public Broadcasting Service. *American Experience*. http://www.shoppbs.pbs.org/wgbh/amex/fever/index.html. Accessed January 15, 2020.

"The Panama Railroad Era." https://cms.uflib.ufl.edu/pcm/timeline/railroad.aspx. Accessed January 20, 2020.

"Trans-Atlantic Slave Trade Database." https://www.slavevoyages.org/. Accessed November 10, 2020.

"When Was the Yellow Fever Epidemic?" https://www.answers.com/Q/When_was_the_yellow_fever_epidemic. Accessed March 4, 2020.

Whipps, H. 2008. "How Sugar Changed the World." https://www.livescience.com/4949-sugar-changed-world.html. Accessed January 22, 2020.

William C. Gorgas. "Chief Sanitary Officer." http://www.kumc.edu/school-of-medicine/history-and-philosophy-of-medicine/clendening-history-of-medicine-library/special-collections/panama-canal/william-c-gorgas-chief-sanitary-officer-in-panama.html. Accessed January 22, 2020.

"Yellow Fever, the Plague of Memphis." http://www.historic-memphis.com/memphis-historic/yellow-fever/yellow-fever.html. Accessed January 8, 2020.

"Yellow Fever in Panama." http://www.kumc.edu/school-of-medicine/history-and-philosophy-of-medicine/clendening-history-of-medicine-library/special-collections/panama-canal/yellow-fever-in-panama.html. Accessed January 22, 2020.

"Yellow Jack in the United States." 1905. *The Daily News*, August 2. Transcribed by T. Easley. http://www.usgennet.org/usa/ar/county/greene/yellowjackus05.htm. Accessed January 11, 2020.

SECTION 5. WESTERN HONEY BEE

Brown, E. 2020. "How the Roman Army Lost a Battle to Bees." https://medium.com/lessons-from-history/how-the-roman-army-lost-a-battle-to-bees-e2cb6dec45d2. Accessed September 29, 2020.

Budnik, R. 2019. "Insect Infantry: Bee Bombs, Scorpion Grenades & Cyber Bugs." War History Online. https://www.warhistoryonline.com/instant-articles/entomological-warfare-the-use.html. Accessed September 29, 2020.

Carr, K. 2017. "Honey Spreads around the World." https://quatr.us/food-2/honey-spreads-around-world-bee-keeping.htm. Accessed September 29, 2020.

Centers for Disease Control and Prevention. 2019. "QuickStats: Number of Deaths from Hornet, Wasp, and Bee Stings, Among Males and Females—National Vital Statistics System, United States, 2000–2017." MMWR Morbidity and Mortality Weekly Report 2019 68: 649. http://dx.doi.org/10.15585/mmwr.mm6829a5.

DeWeerdt, S. 2015. "The Beeline." *Nature* 521: S50–S51. https://doi.org/10.1038/521S50a.

"Honeybee." 2020. Encyclopedia Britannica. https://www.britannica.com/animal/honey-bee. Accessed September 22, 2020.

Horn, T. 2008. Honey Bees: A History. https://topics.blogs.nytimes.com/2008/04/11/honey-bees-a-history/. Accessed September 22, 2020.

Hosler, J. 2018. *The Siege of Acre, 1189–1191: Saladin, Richard the Lionheart, and the Battle that Decided the Third Crusade*. New Haven, CT: Yale University Press.

Krishna, M. T., P. W. Ewan, L. Diwakar, S. R. Durham, A. J. Frew, S. C. Leech, and S. M. Nasser. 2011. "Diagnosis and Management of Hymenoptera Venom Allergy: British Society for Allergy and Clinical Immunology (BSACI) guidelines." *Clinical & Experimental Allergy* 41: 1201–1220.Kritsky, G. 2010. *The Quest for the Perfect Hive: A History of Innovation in Bee Culture*. New York: Oxford University Press.

Kritsky, G. 2015. *The Tears of Re: Beekeeping in Ancient Egypt*. New York: Oxford University Press.

Lockwood, J. A. 2008. *Six-Legged Soldiers: Using Insects as Weapons of War*. New York: Oxford University Press.

Lockwood, J. A. 2019. "Bugs of War: How Insects Have Been Weaponized throughout History." https://www.history.com/news/insects-warfare-beehives-scorpion-bombs. Accessed September 29, 2020.

McAlister, E. 2017. "The Unexpected Pollinator of the Cocoa Tree." https://www.science-friday.com/articles/meet-the-flies-that-pollinate-cocoa-trees/. Accessed October 8, 2020.

McGrath, K. 2016. "The Origin of Muhammad Ali's Most Famous Quote." https://www.bustle.com/articles/164846-the-origin-of-float-like-a-butterfly-sting-like-a-bee-proved-muhammad-alis-greatness-early. Accessed September 29, 2020.

Metcalfe, T., and J. Hsu. 2017. "Beasts in Battle: 15 Amazing Animal Recruits in War." Live Science. https://www.livescience.com/60518-animals-used-in-warfare.html#:~:text=During%20the%20Vietnam%20War%20in%20the%201960s%20 and,dorsata%2C%20along%20the%20trails%20used%20by%20enemy%20. Accessed September 29, 2020.

Miller, M. 2015. Beekeeping may go back to the origins of agriculture, up to 9,000 years ago. https://www.ancient-origins.net/news-history-archaeology/beekeeping-may-go-back-early-years-agriculture-9000-years-ago-004553. Accessed September 22, 2020.

Milner, A. "Honey Bee Origins, Evolution and Diversity." Bee Improvement and Bee Breeders Association. https://bibba.com/honeybee-origins/. Accessed September 22, 2020.

"Mithridates VI." https://en.wikipedia.org/wiki/Mithridates_VI. Accessed September 29, 2020.

National Honey Board. "Market Research." https://www.honey.com/honey-industry/research/market-research. Accessed September 29, 2020.

Norako, L. K. "Richard Coer de Lion." The Crusades Project, A Robbins Library Digital Project. University of Rochester. https://d.lib.rochester.edu/crusades/text/richard-coer-de-lion. Accessed September 29, 2020.

Pariona, A. 2019. "Which Crops and Plants Are Pollinated by Honey Bees?" World Atlas. com. https://www.worldatlas.com/articles/which-crops-plants-are-pollinated-by-honey-bees.html#:~:text=Honey%20bee%20pollination%20is%20considered%20essential%20for%20eight,variety%20of%20agricultural%20climates%2C%20from%20temperate%20to%20tropical. Accessed September 29, 2020.

Pesticide Action Network. "Beyond Pesticides. Economic Value of Commercial Beekeeping." www.beyondpesticides.org/pollinators. Accessed as http://cues.cfans.umn.edu/old/pollinators/pdf-value/EconomicValueCommercialBeekeeping.pdf. Accessed September 29, 2020.

Rabon, J. 2018. "Brit History: Britain's History with the Crusades." https://www.anglotopia.net/british-history/brit-history-britains-history-crusades/#:~:text=Richard%20had%20agreed%20to%20go%20on%20the%20Third,her%20husband%20and%20William%E2%80%99s%20cousin%20Tancred%20seized%20power. Accessed September 29, 2020.

Rickard, J. 2008. "Siege of Eupatoria, c.72–71 B.C." http://www.historyofwar.org/articles/siege_eupatoria.html. Accessed September 29, 2020.

Seeley, T. D. 2019. *The Lives of Bees*. Princeton, NJ: Princeton University Press.

Shahbandeh, M. 2019. "Honey Market Worldwide and in the U.S.—Statistics and Facts." https://www.statista.com/topics/5090/honey-market-worldwide/. Published August 23, 2019. Accessed September 29, 2020.

Sones, B., and R. Sones. 2009. "Strange but True: Giant Asian Honeybees Were Used as a Weapon in Vietnam Conflict." *The Oklahoman*. https://oklahoman.com/article/3357564/strange-but-true-giant-asian-honeybees-were-used-as-weapon-in-vietnam-conflict. Accessed September 29, 2020.

Suson, E. E. 2020. "The History of Beekeeping." https://hankeringforhistory.com/the-history-of-beekeeping/. Accessed September 20, 2020.

Wakefield, A. 2019. "Battle of Tanga." Encyclopedia Britannica. https://www.britannica.com/event/Battle-of-Tanga. Accessed September 29, 2020.

Wallberg, A., F. Han, G. Wellhagen, B. Dahle, M. Kawata, N. Haddad, Z. L. P. Simões, M. H. Allsopp, I. Kandemir, P. De la Rúa, C. W. Pirk, and M. T. Webster. 2014. "A Worldwide Survey of Genome Sequence Variation Provides Insight into the Evolutionary History of the Honeybee *Apis mellifera*." *Nature Genetics* 46: 1081–1088. https://doi.org/10.1038/ng.3077.

Wells, W. 2020. "2020 Almond Pollination Prices." https://www.thebeecorp.com/post/2020-pollination-prices. Accessed September 29, 2020.

SECTION 6. TYING THE SILKEN THREADS

Editors of Encyclopedia Britannica. 2020. "Haitian Revolution." https://www.britannica.com/topic/Haitian-Revolution. Accessed October 17, 2020.

Keys, P. 2014. "Yellow Fever: Napoleon's Most Formidable Opponent." http://www.historiaobscura.com/yellow-fever-napoleons-most-formidable-opponent/. Accessed October 16, 2020.

Knight, J. 2012. "Napoleon Wasn't Defeated by the Russians." https://slate.com/technology/2012/12/napoleon-march-to-russia-in-1812-typhus-spread-by-lice-was-more-powerful-than-tchaikovskys-cannonfire.html. Accessed October 17, 2020.

Knighton, A. 2017. "Disaster in the Desert: Napoleon's Egyptian Campaign." https://www.warhistoryonline.com/napoleon/napoleons-egyptian-campaign.html. Accessed October 17, 2020.

McCarthy, T. 2018. "Haiti—the Slave Who Defeated Napoleon." https://aroundtheworld-in80currencies.com/2018/03/16/haiti-the-slave-who-defeated-napoleon/#:~:text=Toussaint%20continued%20the%20resistance%2C%20and%2C%20by%201803%2C%20Napoleon%2C,and%20had%20him%20executed%20in%20exile%20in%201803. Accessed October 17, 2020.

Moore, Richard. "Napoleon's Egyptian Campaign." Napoleonic Guide. https://www.napoleonguide.com/campaign_egypt.htm. Accessed October 17, 2020.

Napoleon.org. "The Campaign in Egypt." https://www.napoleon.org/en/history-of-the-two-empires/. Accessed October 16, 2020.

Peterson, R. K. D. 1995. "Insects, Disease, and Military History: The Napoleonic Campaigns and Historical Perception." *American Entomologist* 41: 147–160.

Raoult, D., O. Dutour, L. Houhamdi, R. Jankauskas, P.-E. Fournier, Y. Ardagna, M. Drancourt, M. Signoli, V. Dang La, Y. Macia, and G. Aboudharam. 2006. "Evidence for Louse-Transmitted Diseases in Soldiers of Napoleon's Grand Army in Vilnius." *Journal of Infectious Diseases* 193: 112–120.

Rickard, J. 2006. "French Invasion of Egypt, 1798–1801." http://www.historyofwar.org/articles/wars_french_egypt.html. Accessed October 18, 2020.

Walton, G. 2019. "The Importance of Bees to Napoleon Bonaparte." https://www.geriwalton.com/the-importance-of-bees-to-napoleon-bonaparte/. Accessed October 20, 2020.

INDEX

For the benefit of digital users, indexed terms that span two pages (e.g., 52–53) may, on occasion, appear on only one of those pages.

Tables and figures are indicated by *t* and *f* following the page number

Academy of Sciences of Havana, 189
Acanthamoeba castellanii, 106–7
Acre, 227–28, 241
Aedes aegypti, 143, 156, 172–74, 175, 178–80, 186, 189, 195, 196, 201, 242, 243, 246
Afghanistan, 21–23, 26, 27–28, 30, 31*f*, 33–35, 47, 147–48
Afrotrogla, 125–26
Age of Discovery, 54, 79, 82–83, 85, 143, 148
Agramonte, Aristides, 194
Aksum, 80
Alexander the Great, 22–23, 25–27, 30–31, 33–35, 147–48, 226, 227
Alexander, Tsar, 243, 244–45
Alexandria
  Afghanistan, 22–23
  Egypt, 240–41
Ali, Muhammad (Cassius Clay), 227
almonds, 229–31, 230*f*, 237
alpaca, 104
Altai, 21–22
ammonia, 173–74
amoeba, 61, 106–8, 110
An Lushan, General, 38
anaphylactic shock, 220
Anatolia, 25, 227
anemophilic, 215, 216
Angola, 153–54, 155, 156–58
Anoplura, 126

Antheraea yamamai, 47
anthrax, 69, 115–16, 117–18
Anti-Plague Institute, 113–14, 114*f*
Anxi, 28
Apis cerana, 225–26
Apis dorsata, 228, 238–39
Apis mellifera, 217–18, 218*f*, 222, 223, 225, 232
Arab, 18, 26, 38, 43–45, 97, 144, 147–48, 154, 240–41
Arawaks, 185
arbovirus, 112, 169
Aristotle, 226
Ark (Bukhara), 35, 36, 36*f*
Arkansas, 167–68, 169, 176
astrolabe, 45
Australopithecus, 127
Avicenna, 35
Azores, 149

Babylon, 25
Bacillus anthraci, 117–18
Bacot, Arthur W., 68
Bactra, 22–23
Bactrian camel, 22–23, 43, 54, 97
Baghdad, 25, 35, 39, 49, 86–87, 97
Bagram, 22–23
Balboa, 85
Barbados, 160, 183
barklice, 125–26
Basra, 50
batik, 46

Battle
  of Bull Run, First, 193
  of Cer, 135–36
  of Kolubara, 135–36
  of Manila Bay, 198
  of Ravenna, 81–82
  of San Juan Hill, 191, 201
  of Stalingrad, 136–37
  of Talas, 38
  of Tanga, 228
  of Vertières, 243
Bay of Bengal, 43
Beauperthuy, Louis-Daniel, 188, 189–90
bee
  Africanized, 223
  bumble, 217, 219, 227–28
  honey, 216, 217–20, 218*f*, 221–24, 225–
    32, 237, 246
  solitary, 217
bee bole, 227–28
bee dance, 220–21
bee hive, 225, 227–28, 230–31, 230*f*
bee queen caller, 225
bee rustler, 222
beekeeping, 223, 224, 225–26, 228,
  229, 230–31
beeswax, 222, 224–26, 229
Bellevue Hospital Medical School, 192–
  93, 199
Belt and Road Initiative, 237–38
Berber, 151
Berenbaum, May, 12
Beringian land bridge, 104
biological warfare (bioweapons), 87–
  88, 115–18
Black Death, 68, 85–86, 87, 88–89, 96–97,
  103, 108–9
black vomit, 170, 171, 172
Blackburn, Luke Pryor, 171–72
bobak, 86, 113
bombard, 227–28
Bombus, 227–28
Bombyx mandarina, 4–5, 5*f*, 6, 7*f*, 9, 104
Bombyx mori, 4–6, 4*f*, 7*f*, 11, 13, 13*f*, 21,
  41, 47, 57–58, 61, 104
Bonaparte, Napoleon, 240–42,
  243, 244–46
booklice, 125–26
Boophilus microplus, 190
Borodino, 244

Borrelia recurrentis, 136
Bosporus, 81–82
Brazil, 75–76, 104, 144–45, 153–55, 156–
  58, 157*f*, 182–83, 223, 238–39
British Royal Africa Company, 185
brocade, 50
Bronze Age, 65–66
bubo (-oes), 66–68, 83, 88–89, 90–91, 101
Buddha (-ism), 37, 39, 47
Bukhara, 22–23, 31, 35, 36*f*, 36, 87, 97
Bunau-Varilla, Philippe, 205
Butler, General Benjamin, 164

Cabo Blanco, 151
Cabral, Pedro Álvares, 85, 104, 153
cacao, 216
Caffa, 87–88, 115
Caligula, 48
caliph, 38
Cameroon, 156
camp, concentration, 115–16, 136–37
Canary Islands, 149–50
Cannabis, 18
Canon of Medicine, 35
canut, 56
canvas, 18, 45
Cape Bojador, 151
Cape Verde Islands, 149, 152, 153
capitalism, 150
caravanserai, 32, 33*f*, 98–99
caravel, 148, 151
carbon dioxide, 173–74
Carroll, James, 169–70, 194, 196, 197
Carter, Henry Rose, 195, 196, 197
cat worship, 103
catapult, 87–88, 115, 227–28, 241
cattie, 17
Caucasus, 22–23, 31–32, 113, 222
Cave of the Thousand Buddhas, 39
census, 56–57, 168
Census Bureau U.S., 56–57
Centers for Disease Control (U. S.), 220
Chagas Disease, 35–36
champagne, treatment for yellow
  fever, 208–9
Charming Sally, 160–62
chat, 136
chemical weapons, 115–16
Chichén Itzá, 79–80
chikungunya, 169

cholera, 117–19, 120
coccobacillus, 67, 68, 69
cocoon, 3, 6, 11, 13, 14–15, 15*f*, 16*f*,
   27, 47, 72
coffee, 144, 146, 147–48, 185
coffin ships, 133, 134
Columbian Exchange, 229
Columbo, Frank, 93, 94–95, 99, 101–2,
   103, 105, 106
Columbus, Christopher, 74, 85, 153, 182–
   83, 185
compass, 45, 58n.1
Computer Tabulating-Recording
   Company, 56–57
Conolly, Captain Arthur, 36–37
Constantine, Emperor, 81–82
Constantinople (Istanbul), 21–22, 24*f*, 27–
   28, 32–35, 52–53, 54, 81–82, 83–85,
   86–87, 88, 91n.2, 91, 97–98, 108
Copán, 79–80
cotton, 17, 18, 19, 27–28, 37, 46, 49, 55,
   79, 118, 168, 176, 183, 239
Creole cane, 147, 150, 153, 182–83
Crimean Peninsula, 87
crocodile dung, 225
Crusade, Second; Fourth, 53, 54, 147–48,
   227–28, 241
cryptic species, 108
Cuba, 163*f*, 164, 177, 178, 183, 186–87,
   189–90, 191, 193, 194, 195–96, 197,
   198–99, 200–2, 203, 207, 208, 209,
   212, 237, 239
Culebra Cut, 206
Culex, 169, 174–75
   fasciatus, 189
   pipiens, 190
Curtice, Cooper, 190

damask, 53
Danaus plexippus, 107
Dark Ages, 80–81
Davis, Jefferson, 171
DDT, 114, 137–38, 138*f*, 139, 239
de Brueys, Admiral, 241
de Leon, 85
de Lesseps, Ferdinand, 203–4, 205
Dean, William, 196
dengue, 112, 169
Desert
   Gobi, 28, 64, 86–87

Taklamakan, 21–22, 22*f*, 28, 30*f*, 52,
   62, 86–87
Desertas, 149
Dewey, Commodore George, 198
Dias de Novais, Paulo, 153
Dingler, Jules, 205–6
Directory (French), 240
disaster, ecological, 150–51
DNA, 95–96, 104, 110, 146, 225
drone bees, 219, 220, 222, 226, 232
Dubrovnik, 89
dung, crocodile, 225
dungeon, 130
Dunhuang, 22–23, 28, 39
Dutch, 156, 183
Dynasty
   Han, 15, 27, 28, 31–32, 37, 38, 48–
      49, 50–51
   Persian, 25, 26*t*
   Tang, 26*t* , –51, 37, 38, 147
   Yan, 26*t*

ectoparasite, 74, 98, 99, 101, 105, 126
Edict of Nantes, 54
eggs, ostrich, 151
Egypt (Egyptian), 18, 19–20, 21, 25, 26*t*,
      75–76, 83, 154, 224–25, 228, 232,
      239, 240–41
Eiffel, Gustave, 205
Elba, 246
Elfrith, Daniel, 155
Empire
   Arabian, 31, 38
   Byzantine, 26*t*, 38, 49, 50–51, 52–54,
      80, 81–82, 83, 84, 87, 88
   Eastern Roman, 50–51, 80, 81–82
   Gallic, 80
   Napoleonic, 240
   Palmyrene, 80
   Persian, 31–32, 38, 50, 225–26
   Srivijaya, 46
   Timurid, 40–41
   Western Roman, 50–51, 79, 80–83
encephalitis, 112, 169
England, 21, 54, 84–85, 130, 131–32, 147–
      48, 155, 185, 212, 227–28, 240
entomophilic, 215–16
enzootic cycle, 113
epizootic, 92n.3
Ethiopia, 83

Euprojoannisia, 216
eusocial, 218–19
exhumation (-ed), 89, 95–96

Falk, Peter, 93
Famine, Irish, 131–33
Ferdinand and Isabella, 85
Ferghana, 31
ferrets, 113
fever sheds, 134
fibroin, 11
filarial worm, 190
Finlay, Carlos, 177*f*, 177–79, 186–87, 189–
    90, 194–97
flacherie, 55
flavivirus, 169
flax, 18–20
flea, 74, 103–4
    dog, 74, 103–4
    human, 74, 77, 103–4
    Northern rat, 102
    Oriental rat, 73, 73*f*, 75, 91, 101–2,
        105, 108
fomite, 194, 197
Forcipomyia, 216
Forrest City, 168, 169, 176
Forrest, Nathan Bedford, 169, 181n.4
Fort Barrancas, 193, 199
Fort Brown, 199
Fort Detrick, 120
Francis I, King, 54
Frank, Anne and Margot, 136–37
Frankopan, Peter, 41, 42n.2, 42n.3
frass, 129
French Canal Company, 205
French West Indies Corporation, 242
fructose, 144
fruit fly, 239

Gabon, 152–53, 156
gaol fever (jail fever), 131
Gayle, John, 199
Geneva Protocol, 1925, 115, 116, 120
Genghis Khan, 33–35, 39
Genoa, 54, 87
George V., King, 212
gerbil, great, 39–40, 64, 65*f*, 65, 86, 91,
    98–99, 101, 113
gerbil, Mongolian, 64
Gerhard, William Wood, 129–30

germ theory, 54–55, 115, 139, 187, 189
glucose, 144
Goa, 45
Golden Horde, 39–40
Golden Spike, 168
Gonçalves, Antão, 151–52
Gorgas, Lieutenant Josiah, 199
Gorgas, William Crawford, 191–92, 197,
    199–202, 200*f*, 203, 207–12
Gorgas gangs, 208
Gossypium, 18
gram-negative bacterium, 67
Grand Armée, 243–45
Grand Library of Baghdad, 39
Great Buddha Hall, 47
Great Chicago Fire, 166
Great Eastern Temple, 47
Great Epidemic, Russian, 136, 137–38
Great Hunger, Irish, 131–32, 139
Great Plague of London, 89–90
Great Wall, 28
Greenwich, 21
Grey Nuns, 134, 239
Gridley, Captain Charles, 198
Grosse Isle, 133
guinea pigs, 103–4
Gulf of Oman, 43–45
gunpowder, 37, 41, 240
gynaecea, 52–53

Hagia Sophia, 82, 91n.2
Haiti, 74, 160–62, 183, 242, 243, 246
Hamilton, Alexander, 162
haplodiploidy, 219, 233n.7
Harbin, 116–17, 117*f*
Havana, 163*f*, 164, 169–70, 186, 187,
    189, 191–92, 193, 196–97, 199–203,
    211, 212
Havana Yellow Fever Commission, 169–70
Hayes, President Rutherford B., 166
Hazda, 223
hemp, 17, 18–19
Hengduan, 62–64
Henry the Navigator, Prince, 85, 148–
    50, 151
Herat, 22–23
Hexi Corridor, 28
Himalayas, 21–22, 30, 62
Hindu Kush, 21–22, 26*t*, 30, 31*f*,
    43, 62, 97

Hollerith, Herman, 56–58, 57*f*
holometabolism (-olous), 72
holy water, 208
honey hunting, 223–24
Hong Kong, 66, 67, 68, 70, 70*f*, 71, 75, 78, 99
Horn of Africa, 154
horse, Mongolian, 23
Howard Association, 130–31
Howard, John, 130–31
Hsi-Ling-Shih, Empress, 3, 4–5
Huangdi, Emperor, 3–4
Hulagu, 39
hummingbirds, 216
Hundred Year War, 198

IBM, 57–58, 58n.3
Ibn Battuta, 45
ibn Sina (Avicenna), 35
Indonesia, 37, 46, 47, 81
Indus Valley, 18, 43–45, 147–48
Industrial Revolution, 55, 185–86
insecticide, 114, 137, 231–32
Institut, Pasteur, 67, 70, 70*f*, 71, 93–94, 135
Iran, 18, 23, 25, 26*t*, 27–28, 51–52
Iranian Plateau, 18
Iraq, 25, 26*t*, 27–28
Ishii, General Shiro, 116–17, 118–19, 120
Ishii bacterial bomb, 119
Islam, 33–35, 36–37, 38, 46, 53, 80, 109
Istanbul (Constantinople), 21–22, 24*f*, 27–28, 32–35, 52–53, 54, 81–82, 83–85, 86–87, 88, 91n.2, 91, 97–98, 108
Isthmian Canal Commission, 206, 207, 209

Jacquard loom, 55–56, 56*f*, 238–39
Jade Gate, 28
jail fever (gaol fever), 131
Jamaica, 155, 183
Japan, 6, 8–9, 18–19, 27–28, 47, 58n.2, 66, 67, 80, 93–94, 115–17, 118, 119, 120, 137–38, 239
Jefferson, Thomas, 246
Jenner, Edward, 189
Jews, 58n.3, 85, 88–89, 109, 137, 239
John I, King, 148
John VI, Emperor, 88

Johnstown Flood, 166
Jope, John, 155
Justinian I, 51–52, 81–84

K'unming, 66
Kandahar, 22–23
Karakorams, 21–22, 30, 62
Kashgar, 30
kaya, 18–19
Kazakhstan, 21–22, 26*t*, 65*f*
Khotan, Prince of, 52
kissing bug, 35–36, 140n.2
Korea, 6, 8–9, 11, 26*t*, 27–28, 47, 49, 116
Kritsky, Gene, 225
Kublai Khan, 25–27, 39–40
Kunlun Shan, 21–22, 22*f*, 28
Kurgan Horse Culture, 23
Kutuzov, General, 244, 245
Kwantung Army, 116–17, 117*f*
Kyrgyzstan, 21–22, 26*t*, 78

L'Orient, 241
L'Ouverture, Toussaint, 242
lactic acid, 173–74
Lambert, Alexander, 207–8, 210, 212
larva, 11, 12, 13–14, 45, 51, 52, 53, 72, 152, 172, 174, 175, 202, 208, 220
lateen, 148
Lazear, Camp, 196–97
Lazear, Jesse, 194, 195, 196, 197
LeClerc, General Charles, 242–43
Lepidoptera, 5–6
Lesseps, Ferdinand de, 203–4, 205
Lesser Antilles, 160, 184*f*
Lewis, C. S., 136
lice
    body, 35–36, 105, 126, 128–29, 131, 134, 136–37, 139, 244, 245
    chimpanzee, 128
    head, 35–36, 126, 128, 239
    human, 105, 110, 126, 128
    pubic, 126–28, 239
Lincoln, Abraham, 164, 171, 238–39
linen, 17, 18–20
Lister Institute of Preventive Medicine, 75
Little Rock, 163*f*, 168–69
llama, 104
Lockley, Tim, 170, 179
Lockwood, Jeff, 120–21n.2, 227, 233n.14
locust, desert, 239

London School for Hygiene and Tropical Medicine, 212
Louis XI, King, 54
Louisiana Purchase, 246
Luanda, 153–54
Lundgren, Jonathan, 231–32, 233–34n.16
lymph node, 66–67, 70–71, 83
lymphatic system, 66–67, 70–71, 83, 90–91
Lyon, 54, 56, 88

Macaronesia, 149
MacArthur, General Douglas, 120
Madagascar, 66, 75–76, 103, 113, 174
Madeira, 143, 147f, 149–52, 153, 154–55, 158n.4, 182–83
Magellan, 85
Maine, U.S.S., 191
Malacopsyllidae, 72
malaria, 70, 112, 135, 137, 177, 191–92, 199, 203, 205, 207, 208–9, 211–12, 239
Malta, 240, 241
Mamluks, 240–41
Manchukuo, 116–17
Manchuria, 116, 120
Manson, Patrick, 190
Manuel, King, 153
Marco Polo, 25–27, 39, 45
Maritime Routes, 27–28, 30, 37, 41, 43, 44f, 46, 50, 54, 62, 75, 97, 98
marmot, 64, 65, 67, 77, 86, 113
Marshall, James, 204
Martin, Charles J., 75
Martinique, 183
McKinley, William, 193, 198
mead, 226
Mecoptera, 72
medical refugees, 162–64, 165
Medina, 109
Memphis, 162–64, 163f, 165–66, 165f, 167–69
Mendes de Vasconcellos, Luis, 153–54
Merv, 22–23, 86–87, 97
Mesopotamia, 49–50
metallurgy, 37
metamorphosis, 5–6, 14, 222
miasmas, 70–71, 84, 88–89, 90, 189, 193–94
Middle Ages, 25, 54, 79, 80, 82–83, 85

middle class, emergence, 109
Middle Passage, 160, 185–86
midge, chocolate, 216
Mills, John Easton, 134
Milne, A. A., 136
Min (Egyptian god), 225
Mithridates, 227
Moche, 79
Mongolia, 8–9, 16, 23, 26t, 37, 64, 85, 86–87, 93–94, 113, 116
monsoons, 37, 46
More, Thomas, 130
Morgan, Thomas Hunt, 239
Muhammad, 38, 109, 227
mulberry, white, 3–4, 5, 11, 12–14, 13f, 17, 43, 46, 47, 51, 52, 54, 55, 61
multi-generational migration, 107
Mumbai, 66, 70, 100
Muridae, 64
Muscat, 45
Muscovy duck, 104
mutation, 7–8, 96
mutualism, 145, 216–17

Nannochoristidae, 72
Naples, 137
Nara, 47
Natchez, 171
National Honey Board (U. S.), 229
Nazca, 79
Nazrullah Khan, 35–36
Ndongo, 153
nectar, 174–75, 216–17, 220, 221–22, 221f, 231–32
Nelson, Horatio, 241
Neotrogla, 125–26
Nestorian monks, 51–52
nettle, 19
New Guinea, 46, 145, 146
New Orleans, 77, 162–64, 165–66, 186, 195, 246
New York, 134, 137, 160–62, 171, 192–93, 199, 204
Newton, Sir Isaac, 35, 90–91, 92n.4, 239
Nicolle, Charles, 135
Nika riots, 82
Ningbo, 118
nit, 125
Nobel Prize, 62, 135, 179–80, 220, 239
Norfolk, 156, 162, 166

Nosema bombycis, 55
Nott, Josiah, 188, 189–90

octenol, 173–74
Odoacer, King, 81–82
Olivarius, Kathryn, 180
oviposit(-ion), 55, 173f, 174–75
ovipositor, 219–20

Pacific Theater, 239
Pakistan, 18, 27–28, 71
Pamirs, 21–22, 22f, 30, 62
Panama Canal, 182, 186–87, 197, 200f, 209f, 210f, 211f, 212, 237
Pandemic, first, 83, 86, 88, 89, 102, 103–4, 108–9
Pandemic, Justinian (second), 83, 84, 85, 88, 89, 96–98, 100, 102, 103–4, 108–9
Pandemic, third, 65–66, 75–76, 77–78, 86, 92n.3, 94–95, 96, 98, 99, 100, 104, 112–13
paper, 12, 18–19, 37, 38, 41
Parthian, 26t, 31–32, 38, 48–49, 50–51
Pasteur, Louis, 54–55, 189, 238–39
pasteurization, 55
Pax Mongolica, 39
pebrine, 55
Pediculus humanus, 105, 126, 127f, 128, 139
Pediculus schaeffi, 128
Persepolis, 25
Persia, 49, 50, 53, 80, 84, 147–48, 225–26
Persian Gulf, 43–45, 97
pestilence, 85–86, 109–10
pestis secunda, 89
phagocytosis, 106–7
pharate, 72
Philadelphia, 129–30, 134, 160–62, 180–81n.1, 187
Philippines, 198–99
Phthiraptera, 126
Phthirus gorillae, 126–27
Phthirus pubis, 126
Phytophthora infestans, 131
pikas, 113
Pingfan, 117–18, 117f, 120
Placerita Canyon, 204

plague
    bubonic, 39–40, 65–67, 82–83, 85, 93–94, 107–8, 116, 129, 139
    pneumonic, 66, 83, 93–94
    septicemic, 66, 93–94, 102
Plague of Athens, 129
Plasmodium, 112
pollen basket, 218f, 221–22
pollination, 215, 216–17, 229–30, 230f, 231, 232
Pontus, Kingdom of, 227
Porto Santo, 149
Portugal (-uese), 66, 80–81, 85, 104, 148–50, 151–55, 182–83
potatoes, 131–32, 140n.3
prairie dog, 77, 86, 113
Prìncipe, 152–53
prison reform, 130–31
prisons, British, 130–31
privateer, 155–56
Procopius, 83–84
Prowazek, Stanislaus von, 135
Psocodea, 125–26
Puerto Rico, 183, 186, 198–99
Pulex irritans, 77, 103–4
pupa, 5–6, 12, 14–15, 72, 222
pupation, 14, 51
Pyramids, Giza (Egypt), 240–41
pyrethrum, 166, 202, 208

Qinling-Huaihe Line, 8–9
Quanzhou, 45
quarantina giorni, 164, 181n.2
quarantine, 89, 133–34, 164, 167–68, 170, 176, 192, 194, 195, 238–39
Quran, 35

Ragusa, 89
railroads, 167–68
    Chinese Eastern, 116
    Memphis & Little Rock, 160–62, 163f
    Panama Railroad Company, 204
rain barrels, 202
ramie, 17, 19
rat
    Asian house (*Rattus tanezumi*), 66, 94, 99, 118
    black (*Rattus rattus*), 75, 94, 100
    brown (*Rattus norvegicus*), 64, 65, 75, 91, 94, 99–100, 101, 102, 106

Re (Egyptian sun god), 224
Rebellion, An Shi (An Lushan), 38
Reed, Walter, 177, 178f, 182, 187, 192–93,
     194, 195, 196–97, 200–1, 207, 211,
     212, 237
relapsing fever, 129, 136
Renaissance, 109
Revolt
   French silk workers, 56, 238–39
   Haitian Slave, 160–62, 242
   women's silk ban, 238–39
Revolution
   Chinese Agricultural, 9–10
   Industrial, 55, 185–86
   Neolithic Agricultural, 9
   Secondary Products, 10
Ricketts, Howard, 135
Rickettsia prowazekii, 112, 129, 135, 136,
     139, 244
River
   Amu Darya, 30
   Duong, 43
   Euphrates, 25, 97
   Hotan, 28, 62–64
   Indus, 38, 43–45, 97, 147–48
   Mekong, 62–64
   Mississippi, 162–65, 163f, 167–68,
     195, 246
   Nile, 203
   Salween, 62–64
   St. Lawrence, 133, 134
   Tarim, 28
   Tigris, 25, 39, 97
   Yangtze, 62–64
Road (-s)
   Fur, 97, 99–100
   Royal, 25–27
   Tea-Horse, 62
Rocha Lima, Henrique da, 135
Rochalimaea quintana, 136
Rochambeau, General, 242–43
Romulus Augustus, 80
Roosevelt
   Franklin, 180–81n.1
   Theodore, 191, 192f, 197, 198, 199, 206,
     207–8, 210, 210f, 238–39
Rothschild
   Charles, 73–74, 75, 101–2
   Lionel, 73–74
   Miriam, 73–74, 78n.2
Rough Riders, 191, 192f, 194, 199

roundel, 50
Route, Maritime, 30, 50
Route, Spice, 48–49
rum, 160, 185–86
Rush, Benjamin, 162
Russia, 6, 27–28, 82–83, 86, 87, 97, 98,
     108, 116, 136, 137–38, 227, 243–
     45, 245f
Russian Army, 243, 245

Saccharum
   barberi, 145
   officinarum, 145, 147
   robustum, 145, 146
   spontaneum, 145
Saint-Domingue, 183, 242, 246
saliva, 14, 23–25, 101
salivary, (glands), 11, 14, 32–33, 41, 46,
     57–58, 61, 156–58
Salmon, Daniel, 190
Salmonella, 102
Salmonella enterica, 117–18
Samarkand, 22–23, 31, 33–35, 34f, 40,
     40f, 86–87, 97
samite, 53
San Francisco earthquake, 166
San Juan Heights, 191, 201
San Juan Hill, 191
Santa Clarita Valley, 204
Santiago (Cuba), 191, 201
São João Bautista, 155
São Tomé, 152–53, 156–58, 182
Sasanian, 26t, 38, 50–51, 82–83, 84
savanna cycle, 172
Savannah, 162, 170, 179
Scythians, 25, 26t, 47
Sea
   Adriatic, 89
   Andaman, 62–64
   Arabian, 26t, 30, 43, 97, 148
   Aral, 26t
   Black, 23, 26t, 28, 53, 80, 81–82, 87–88,
     97–98, 106, 148, 227
   Caribbean, 156–58, 160–62, 164, 182–
     83, 184f, 185–86, 191, 204, 242
   Caspian, 26t, 28, 87, 97–98
   China, 43–45, 62–64, 99
   of Japan, 47
   Marmara, 81–82
   Mediterranean, 80, 106, 203–4
   Red, 43–45, 86–87, 97, 198

Seal, Mediterranean monk, 143, 151
seaport, Mediterranean, 100
Seeley, Tom, 223–24, 231, 232–33n.4
selection, natural, 8, 9
Seneca the Younger, 48
Serbia, 135–36
sericin, 11
sericulture, 3–4, 12, 13, 46, 51–52, 54
Shanghai, 62–64, 99
shanyu, 16, 17
Shibasaburo, Kitasato, 67, 69*f*, 93–94
ship fever, 133
shipworms, 45
Shonts, Theodore, 209–11
silk, 3–20, 21–42, 43–58
  Byzantine, 53
  Persian, 49*f*, 50
  tensile strength of, 11
silk moths
  domestic (Bombycidae), 47
  giant (Saturniidae), 47
  Japanese, 47
Simond, Paul-Louis, 71, 75, 86,
    100, 101–2
Siphonaptera, 72
skep, 224
Slav, 154
Slave Coast, 157*f*, 185
Slave Trade Act, 186
Slavery Abolition Act, 183
Smith, Theobald, 190
Sogdiana, 30–31
Sogdians, 30–31, 37
South African Sugarcane Research
    Institute, 145
Spanish
  army, 191
  colonies, 186
  court, 85
Spice Islands, 37
steppe, 16, 23–27, 31, 39–40, 47, 53, 86,
    97, 99, 108
Steppe Roads, 23–25, 27, 28
Sternberg, George, 187, 188–89, 188*f*,
    190, 192, 193–95, 197
Stevens, John F., 206–7, 208
Stevens, Nettie, 239
Stoddart, Colonel Charles, 36–37
Strait
  Malacca, 53
  Sunda, 53

Stranger's Disease, 164
sucrose, 144
Suez Canal, 203–4, 205
sugar, 143–59
sugar beet, 144–45
sugar consumption, 185
sugarcane, 143–59
Sumatra, 46, 143, 145–46
Surgeon General, U. S., 187, 188*f*, 192,
    193–94, 195, 200*f*, 207, 212
Sutter's Mill, 204
Swahili, 80
synanthropy, 64, 94
Syria, 18, 26*t*, 240, 241

Tabulating Machine Company, 56–57
Taizong, Emperor, 147
Tajikistan, 21–22
tarbagan, 86
Tarim Basin, 21–22, 22*f*, 26*t*, 28, 30*f*, 30–
    31, 38, 62, 64, 86
Taxila, 30
tectonic plates, 21–22
teeth, 89, 95–96, 97, 110n.1, 237
Tel Rehov, 225
Teotihuacan, 79–80
Texas cattle fever, 190
Theodora, 82
Three Parallel Rivers Protected
    Area, 64
Tibetan Plateau, 6, 8–9, 21–22,
    62–64, 63*f*
Tien Shan, 21–22, 22*f*, 28, 62, 78, 97
Tikal, 79–80
Timur (Tamerlane), 33–35, 40*f*, 40
toad vomit, 90–91, 92n.4, 225, 239
Toba volcano, 145–46, 158n.2
Tolkein, J.R.R., 136
Toulon, 240
Trade Triangle, 185–86
trade winds, 148, 153
Treasurer, 155–56
Treaty of Hidalgo, 204
Treaty of Paris, 198–99
tree beekeeping, 224
tree hole mosquitos, 172, 174–76
trench fever, 129, 136, 245
Tristão, Nuno, 151
trophallaxis, 218–19
Trypanosoma, 35–36
Tunga penetrans, 74

Turkey, 26t, 27–28, 41–42n.1, 222, 225, 227
turkey, 104
Turkmenistan, 21–22, 26t
twill, 50, 53
typhoid fever, 117–18, 120, 129–30, 134, 191–92, 193–94, 199, 200, 202
typhus, 125–40
  epidemic, 129, 135
  murine, 129
  scrub, 129

Ulugh Beg, 33–35
Ulugh Beg Observatory, 33–35, 34f
Unit 731, 116–18, 117f, 119, 120
United States Army
  Medical Corps, 193
  Medical School, 193
  Typhoid Board, 193–94
  Yellow Fever Commission, 192
University of the South, 199
unwashed masses, 132
Upper Silesia, 138–39
urea, 173–74
Uzbekistan, 21–22, 27–28, 34f, 35, 36f, 40f, 49f

vacuoles, 106–7
Valletta, 240
Venice, 54, 89, 164
Vespa mandarina, 205
Vibrio cholerae, 117–18
Vicksburg, 165
Vietnam, 26t, 43–45, 62–64, 68, 70, 228, 238–39
Vilnius, 243, 244–45
Virchow, Rudolf, 138–39
volcanic eruptions, 81, 82, 84, 98–99
Volta do Mar, 153
von Frisch, Karl, 220–21
von Richtofen
  Ferdinand, 27
  Manfred, 27

waggle dance, 220–21, 233n.10
Walker, John C., 209, 210–11
Wallace, John Findley, 206
War
  Civil, Byzantine, 88

Civil, U. S., 134, 164, 168, 169, 171, 193, 199, 204
Gothic, 84
Lazic, 84
Revolutionary, American, 186
Second Sino-Japanese War, 116
Spanish-American, 192f, 198, 199, 203, 206
Warsaw Ghetto, 137, 140n.6
Washington, George and Martha, 162
Watson, Thomas J., 57–58
Welwitschia, 232n.3
West Nile fever, 112, 169
wheeled transportation, 37
White Gold, 183, 197, 204
White Lion, 155–56
white man's flies, 228
Wilder, Russel, 135
Wilson, Woodrow, 212
Wood, Leonard, 199, 201
woodwasp, 219–20
wool, 17–18, 19, 49
World Health Organization, 77, 113
World War I, 113, 115–16, 135–36, 228
World War II, 62–64, 90, 114, 115–16, 136–39, 176, 198, 239
Wu, Emperor, 12

Xenopsylla cheopis (Oriental rat flea), 73, 73f, 75, 91, 101–2, 105, 108
Xi'an, 27–28, 32–33
Xiongnu, 16–17, 26t, 27, 31–32
Xiyu, 16–17

Yellow Jack, 165, 170, 203
Yersin, Alexandre, 67, 68f, 68–71, 70f, 86, 93–94, 100
Yersinia pseudotuberculosis, 96
Yucatan, 155, 156–58, 160, 175–76, 238–39
Yuki-Tsumugi, 47
Yunnan, 43, 62–64, 65–66, 78, 94, 119

Zanzibar, 45
Zika, 169
Zimbabwe, 80
zoonosis (zoonoses), 65, 107–8
zoonotic, 65–66, 74, 86, 92n.3, 94–95, 112, 113